MW00844078

Problem Books in Mathematics

Edited by P.R. Halmos

David W. Cohen

An Introduction to Hilbert Space and Quantum Logic

With 38 Illustrations

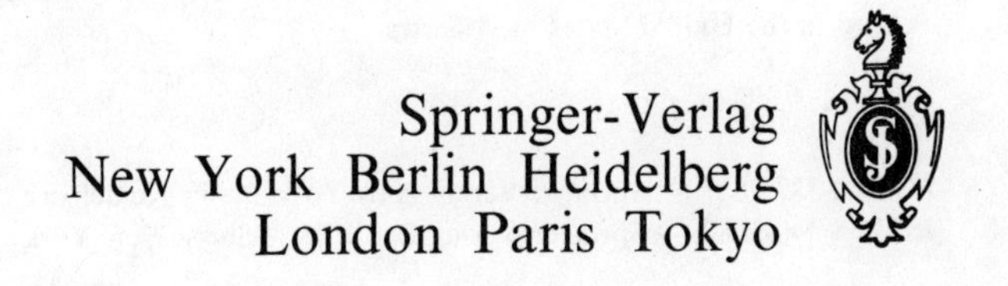

David W. Cohen
Department of Mathematics
Smith College
Northampton, Massachusetts 01063
U.S.A.

Series Editor
Paul R. Halmos
Department of Mathematics
University of Santa Clara
Santa Clara, California 95053
U.S.A.

AMS Subject Classification: 46C, 81, 81A12, 47B15, 47A25, 35P05

Library of Congress Cataloging-in-Publication Data
Cohen, David W.
An introduction to Hilbert space and quantum logic/David W. Cohen
p. cm.—(Problem books in mathematics)
Bibliography: p.
1. Hilbert space. 2. Quantum theory. 3. Logic, Symbolic and mathematical. I. Title. II. Series.
QA322.4.C64 1989
515.7′33—dc19 88-24989

Printed on acid-free paper

Phototypesetting by Thomson Press (India) Ltd, New Delhi, India.
Printed and bound by R.H. Donnelley & Sons, Harrisonburg, Virginia.
Printed in the United States or America.

9 8 7 6 5 4 3 2 1

ISBN 0-387-96870-9 Springer-Verlag New York Berlin Heidelberg
ISBN 3-540-96879-9 Springer-Verlag Berlin Heidelberg New York

Dedicated to
My parents, Rose and Lew
My wife, Doris (mitt trygt punkt),
My children, Bonnie and Sara
My sister, Carol

Preface

Content

The often-lamented communications gap between mathematicians and theoretical physicists begins to form at the undergraduate level. Some undergraduate physics majors learn little mathematics beyond linear algebra and differential equations, while mathematics majors often see *no* modern physics. The result of this gap is that some of the most profound and mathematically interesting questions about the foundations of quantum physics are accessible only to those few researchers who pursue mathematics and physics at a very advanced level.

This book was written to bridge the gap at a lower level. It should be accessible to undergraduate and beginning graduate students in both mathematics and physics. The only strict prerequisites are calculus and linear algebra, but the level of mathematical sophistication assumes at least one or two intermediate courses—for example, in mathematical analysis or advanced calculus. No background in physics is assumed. Historically, nonclassical physics developed in three stages. First came a collection of ad hoc assumptions and then a cookbook of equations known as "quantum mechanics." The equations and their philosophical underpinnings were then collected into a model based on the mathematics of Hilbert space. From the Hilbert space model came the abstraction of "quantum logics." This book explores all three stages, but not in historical order. Instead, in an effort to illustrate how physics and abstract mathematics influence each other, we hop back and fourth between a purely mathematical development of Hilbert space and a physically motivated definition of a logic, partially linking the two throughout, and then bring them together at the deepest level in the last two chapters.

To explore the interplay between mathematics and physics I was forced to make choices not only in the selection but also in the treatment of some topics. For example, the opening chapter on integrals with respect to measures on the Borel sets is rather specialized. Topics important to a general development of integrals are replaced by those more important to quantum physics such as the idea of integrating the identity function on the reals with respect to a complex-valued measure. Similarly, the concluding chapter on quantum mechanics presents the Schroedinger equation with practically no physical motivation in order to relate it quickly to the general notion of observables and their mathematical connection to Hilbert space logic.

To be reasonably comprehensive I have included some mathematics whose proof is beyond the level of this book. The insistence that no result may be used until its proof has been digested (or at least presented) may breed healthy skepticism in students, but it also keeps their mathematical horizons rather narrow. Researchers, on the other hand, frequently use the results of others without combing through the proofs. I see no reason to withhold from a reader a lovely presentation of the spectral theorem for bounded Hermitian operators, for example, as long as the reader can understand the statement of the theorem—even if its proof is well beyond the level of this presentation.

It is my hope that by combining some deep mathematics with some deep physics at a level that is not formidable this book can encourage a reestablishment of links across the communications gap that has been steadily widening recently—to the detriment of both mathematics and physics.

Format

Nearly all of the proofs and many of the examples are in the form of projects for the reader to complete. The coaching manual, which follows the main text, contains hints for completing the easier projects, but mostly it contains complete solutions.

There are two reasons why the book is written with emphasis on projects. One is so that a reader who desires merely an acquaintance with this material can get that by reading only the main text. The other is to provide the more thorough reader with the opportunity to get deeply into the material by proving the theorems or by independently investigating important examples. Some of the projects consist of routine exercises, but many (marked by *) require some sophisticated ideas, while some (marked by **) I consider really too difficult for most readers to complete without consulting the Coaching Manual. I believe there is great value in writing out a solution to a difficult project, even if most of the ideas come from the Coaching Manual.

Use

The format allows using the book for independent study or as a text in a lecture course. Naturally, an instructor can choose a blend of lecturing about some projects and assigning others as independent work, depending on how advanced the students are and how much is to be covered in the time available. A thorough study of all the topics and completion of all the projects is likely to require two semesters for very advanced undergraduates or beginning graduate students. On the other hand, I have successfully taught two one-semester courses at Smith College based on this material.

In an intermediate level course (junior and senior math, physics, and chemistry majors) I covered Chapters 3, 4, and 6A. I spent most of the time on Chapter 3, stressed the finite dimensional cases in Chapter 4, and lectured on the main ideas in 6A, filling in with material on measure and integration (Chapter 1) on a "when needed" basis. The projects I assigned were among the easier ones.

For an advanced course (our best senior math and physics majors) I covered Chapters 2 through 6, assigning about half the projects for independent work.

Acknowledgment. It is with the greatest pleasure that I acknowledge the roles of the following people who have helped to make this book possible. David Foulis and the late Charles Randall got me into this business and provided constant encouragement and mathematical support. Gottfried Rüttimann graciously arranged for me and my family to spend a year in Bern, where under his guidance I found an entire world of fascinating connections between mathematics and physics. George Svetlichny and James Henle collaborated with me in research projects that provided valuable stimulation and insight into mathematical and philosophical questions. Arthur Swift spent hours helping me sort out the ideas in Chapter 8. Elizabeth Kumm carefully read every sentence of the manuscript with a critical eye and was responsible for dozens of significant improvements. Marjorie Senechal read portions of the manuscript and provided valuable suggestions and encouragement. Kathy Zaffiro patiently did all the artwork.

One might hope that with all this support I have written a book without errors. Alas, dear reader, I suspect that this is not the case, and I alone bear the responsibility for the deficiencies that remain.

David W. Cohen

Symbols Used but Not Defined in the Text

$\mathbb{N}$	the set of natural numbers.
$\mathfrak{R}$	the set of real numbers.
$\mathfrak{R}^\infty$	the extended real numbers: $\mathfrak{R} \cup \{\infty\}$.
$\mathfrak{C}$	the set of complex numbers.
$[a, b)$	$\{x \mid x \in \mathfrak{R}$ and $a \leq x < b\}$.
$\varnothing$	the empty set.
$A \backslash B$	$\{x \mid x \in A$ and $x \notin B\}$ (A and B are any sets).
$\cup \mathbb{S}$	the union of the sets in $\mathbb{S}$: $\{x \mid x \in S$ for some $S \in \mathbb{S}\}$. If $\mathbb{S}$ is indexed by $\mathbb{N}$, we also write $\bigcup_n S_n$ or $\bigcup_{n=1}^{\infty} S_n$ for $\cup \mathbb{S}$.
$\cap \mathbb{S}$	the intersection of the sets in $\mathbb{S}$: $\{x \mid x \in S$ for every $S \in \mathbb{S}\}$. If $\mathbb{S}$ is indexed by $\mathbb{N}$, we also write $\bigcap_n S_n$ or $\bigcap_{n=1}^{\infty} S_n$ for $\cap \mathbb{S}$.
$\sup(T)$	the least upper bound of the set T, if T is a set of real numbers bounded above, or ∞ if T is a set of numbers not bounded above.
$\langle\langle x_n \rangle\rangle$	a sequence indexed by the natural numbers. For additional clarity we sometimes write $\langle\langle x_n \rangle\rangle$ $(n \in \mathbb{N})$ to denote a sequence indexed by the natural numbers.
$\lim_n \langle\langle x_n \rangle\rangle$	the limit of the sequence $\langle\langle x_n \rangle\rangle$. For additional clarity we sometimes write $\lim_{n\to\infty} \langle\langle x_n \rangle\rangle$ for this limit.
$f(x)$	the value of the function f at domain element x.
$f[A]$	$\{f(x) \mid x \in A\}$.
$f^{\leftarrow}[B]$	$\{x \mid f(x) \in B\}$.

Contents

CHAPTER 1

Experiments, Measure and Integration

Introduction

We learn about our physical universe by doing experiments. That is, first we do something such as flip a coin, or touch a hot stove, or measure how long it takes a marble to drop from a certain height. Then we record what happens after we do it—the coin comes up heads, we get burned, the marble takes 6 seconds to drop. What we record is called an outcome of the experiment. We identify an experiment by its outcome set, so we can write $C = \{\text{heads}, \text{tails}\}$ to denote the coin flip experiment.

An experiment is most useful if it is repeatable and its outcomes are describable precisely. For example, we can flip a coin many times and each time we can record heads or tails precisely. We can repeatedly drop a marble, but the set of outcomes depends on how precisely we wish to measure the time. If we require accuracy only to 1 second, the outcome set could be $M = \{5, 6, 7\}$, where the numbers denote seconds. If we require accuracy to one decimal place, we might have an outcome set with 21 numbers in it: $M = \{5.0, 5.1, \ldots, 6.8, 6.9, 7.0\}$. In an experiment where we touch a hot stove it is a little difficult to describe the outcome with precision. Did we get "badly burned" or only "slightly burned"? A numerical measurement of temperature or a count of the number of skin cells damaged might serve as a useful outcome set.

In many experiments the outcomes consist of a set of real numbers. In fact, it is sometimes convenient to introduce numbers artificially, as, for example, by assigning the number one to the outcome "heads" and the number zero to "tails." We begin our study of experiments, therefore, by considering the real number system.

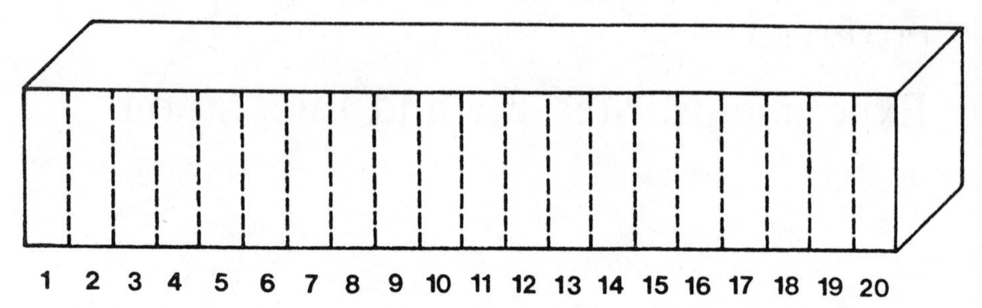

Figure 1A.1. A physical system consisting of a box with a marked window and a firefly inside.

Part A: Measures

Consider the following physical system. Let us take a firefly and put it into a box. The front of the box is a clear plastic window with thin vertical lines drawn on it to divide the window into 20 equal-sized sections. (See Figure 1A.1.)

We shall define a simple experiment on our physical system: Look through the window. The outcomes of this experiment will be numbers: 0 if we see no light when we look, or a number n between 1 and 20 if we see a light in section n of the window. We denote the experiment by $E = \{0, 1, 2, \ldots, 20\}$.

Suppose we perform the experiment p times for some large number p. Each outcome $x \in E$ can be associated with a "weight" $\omega(x) = m/p$, where m is the number of times x occurred. Then $\omega(x)$ is the proportion of times that outcome x occurred among all the performances of experiment E. We can think of $\omega(x)$ as the weight of x in E. This motivates the following definition.

1A.1. Definition. A *weight function* on a finite set E is a function $\omega: E \to [0, 1]$ with the property $\sum_{x \in E} \omega(x) = 1$.

Notice that E need not be a set of numbers, although that is what it is in our example above. If E is a set if numbers, then we can compute a "weighted average" of the numbers as follows.

1A.2. Definition. The *expected value* of a finite set of numbers E with respect to weight function ω is

$$\underline{\operatorname{Exp}(E, \omega)} = \sum_{x \in E} x\omega(x).$$

1A.3. Project. Suppose that the firefly experiment was performed 300 times, and that the outcomes were recorded in the following table.

Outcome $x \to$	0	1	2	3	4	5	6	7	8	9	10	11	12	13	14	15	16	17	18	19	20
No. of $\to$ Occurrences of x	5	8	12	11	11	8	0	14	6	9	11	10	12	45	34	24	52	16	6	2	4

For each $x \in E$ define $\omega(x) = m/300$, where m is the number of times outcome x occurred according to the table. Compute $\operatorname{Exp}(E, \omega)$.

If ω is a weight function on an experiment E, then it is common to call $\omega(x)$ the "*probability of obtaining outcome* x in any performance of experiment E when the system is governed by ω."

Now we might ask: If we perform E when the system is governed by ω, what is the probability of obtaining a number between 5 and 8.5 as the outcome? In other words, what is the relative weight of interval $[5, 8.5]$ with respect to ω? The answer is $\sum_{x \in [5, 8.5]} \omega(x)$. In fact, we can assign a relative weight to *every* subset S of real numbers by

$$\mu(S) = \begin{cases} \sum_{x \in S} \omega(x) & \text{if } S \cap E \neq \varnothing, \\ 0 & \text{if } S \cap E = \varnothing. \end{cases} \tag{1A.1}$$

We think of $\mu(S)$ as the measure of the importance of S with respect to ω, and we call μ a "measure function" for $\mathfrak{R}$, the set of real numbers. Notice that a new weight function will immediately give us a new measure function. Next we shall see how we can compute $\operatorname{Exp}(E, \omega)$ by using the measure function μ. This will be a key idea that we will generalize to arrive at a mathematical formulation of the link between physical experiments, weight functions, and measure functions.

We continue discussing the firefly system above with $E = \{0, 1, 2, \ldots, 20\}$. For $k = 0, 1, 2, \ldots, 40$ consider real number intervals of the form

$$J_k = \begin{cases} \{k/2\} & \text{if } k \text{ is even,} \\ \left(\dfrac{k-1}{2}, \dfrac{k+1}{2}\right) & \text{if } k \text{ is odd.} \end{cases}$$

Notice that $[0, 20] = \bigcup_{k=0}^{40} J_k$. Now if we let $x_k = k/2$ for $k = 0, 1, \ldots, 40$, we have, using Equation (1A.1) above,

$$\operatorname{Exp}(E, \omega) = \sum_{k=0}^{40} x_k \mu(J_k). \tag{1A.2}$$

1A.4. Project. Verify equation (1A.2). Notice that equation (1A.2) holds for every weight ω.

Equation (1A.2) is the key motivational idea for the rest of this chapter. We shall define general measure functions and consider sums such as in equation (1A.2) as approximations to expected values for experiments that might have infinitely many outcomes. Our generalization of the sum will be an integral.

1A.5. Definition. Let X be a set, and let $\mathbb{A}$ be a collection of subsets of X satisfying:

(i) $X \in \mathbb{A}$;
(ii) if $S \in \mathbb{A}$, then $X \setminus S \in \mathbb{A}$;
(iii) if $\mathbb{S}$ is a countable subset of $\mathbb{A}$, then $\cup \mathbb{S} \in \mathbb{A}$.

Then we define a *measure on* $\mathbb{A}$ *for set* X as a function μ satisfying the following:

(i′) $\mu : \mathbb{A} \to \mathfrak{R}^\infty$;
(ii′) $\mu(A) \geqq 0$ for all $A \in \mathbb{A}$, and $\mu(A) < \infty$ for at least one $A \in \mathbb{A}$;
(iii′) if $\mathbb{S} \subseteq \mathbb{A}$ is a pairwise disjoint, countable collection, then $\mu(\cup \mathbb{S}) = \sum_{S \in \mathbb{S}} \mu(S)$.

By the equality in (iii′) we mean either that the left side is a finite number, in which case the series (of nonnegative numbers) on the right converges to it, or that the left side is ∞, in which case the series on the right diverges to ∞.

The reason for specifying what the domain $\mathbb{A}$ of a measure should look like is that there are many important measures for the set $\mathfrak{R}$ of real numbers that do not have all the subsets of $\mathfrak{R}$ in their domains. A collection of sets $\mathbb{A}$ for X satisfying (i)–(iii) is called a *σ-algebra for* X, or a *Boolean algebra*, named for the mathematician George Boole. We sometimes use the phrase *μ is a measure for X* to mean there is some σ-algebra $\mathbb{A}$ in X with μ a measure on $\mathbb{A}$. The members of $\mathbb{A}$ are called the *μ-measurable* sets in X.

1A.6. Lemmas.

A. Show $\mu(\varnothing) = 0$.
B. Let $\mathbb{A}$ *be a σ-algebra for set* X *and* μ *a measure on* $\mathbb{A}$.

(i) *If* $A, B \in \mathbb{A}$ *with* $A \subseteq B$, *then* $\mu(A) \leqq \mu(B)$.
(ii) *If* $\langle\langle A_n \rangle\rangle$ *is a sequence of sets in* $\mathbb{A}$ *with* $A_n \subseteq A_{n+1}$ *for all* $n \in \mathbb{N}$, *then* $\mu(\bigcup_n A_n) = \sup_n \mu(A_n)$.
(iii) *If* $\langle\langle A_n \rangle\rangle$ *is a sequence of sets in* $\mathbb{A}$ *with* $A_{n+1} \subseteq A_n$ *for all* $n \in \mathbb{N}$, *and* $\mu(A_n) < \infty$ *for at least one n, then the sequence* $\langle\langle \mu(A_n) \rangle\rangle$ *is monotone decreasing and converges to* $\mu(\bigcap_n A_n)$.

C. If $\mathbb{A}$ *is a σ-algebra for set* X *and* μ_1, μ_2 *are measures on* $\mathbb{A}$, *then*

(i) $\mu_1 + \mu_2$ *is a measure on* $\mathbb{A}$. [*We define* $(\mu_1 + \mu_2)(A) = \mu_1(A) + \mu_2(A)$ *for all* $A \in \mathbb{A}$.]
(ii) *if* $t \in \mathfrak{R}$ *with* $t > 0$, *then* $t\mu_1$ *is a measure on* $\mathbb{A}$. [*We define* $(t\mu_1)(A) = t(\mu_1(A))$ *for all* $A \in \mathbb{A}$.]

PROOF. Project. □

There is one measure for the reals that is of central importance to a vast amount of physics and mathematics. We call it *Lebesgue measure*, and it is the one that is the link between the integration that one studies in beginning calculus and the more general integrals we shall study later in this chapter. We discuss Lebesgue measure by first describing the "Borel sets" in $\mathfrak{R}$.

1A.7. Definition. Let $\mathbb{I} = \{(a,b] \mid a, b \in \mathfrak{R}\}$. We define the collection $\mathbb{B}$ of *Borel sets in* $\mathfrak{R}$ as the smallest collection of subsets of $\mathfrak{R}$ satisfying:

(i) $\mathbb{I} \subseteq \mathbb{B}$;
(ii) if $B \in \mathbb{B}$, then $\mathfrak{R} \backslash B \in \mathbb{B}$;
(iii) $\mathbb{B}$ is closed under countable unions.

Note that $\mathbb{B}$ must also be closed under countable intersections, because if $\mathbb{S}$ is a countable collection in $\mathbb{B}$, then $\cap \mathbb{S} = \mathfrak{R} \backslash (\cup_{S \in \mathbb{S}} \mathfrak{R} \backslash S)$.

1A.8. Lemmas.

A. If $a, b \in \mathfrak{R}$ with $a < b$, then all of the following are Borel sets:

$$(-\infty, a), (-\infty, a], (a, b), [a, b], \mathfrak{R}, \varnothing$$
$$[a, b), \{a\}, [b, \infty), (b, \infty), (a, b].$$

B. The collection $\mathbb{B}$ of Borel sets is a σ-algebra in $\mathfrak{R}$.

PROOF. Project. □

Lemma 1A.8 can be used to prove that most subsets of the reals that we use in calculus are Borel sets. In fact, the proof that there exists a set that is not a Borel set is rather difficult, and we shall not concern ourselves with it. We turn instead to our main reason for considering $\mathbb{B}$, namely to define Lebesgue measure. The proof of the following theorem is complicated, and we shall omit it.

1A.9. Theorem. *There exist a σ-algebra $\mathbb{A}$ for $\mathfrak{R}$ with $\mathbb{B} \subseteq \mathbb{A}$ and a measure μ on $\mathbb{A}$, called Lebesgue measure, such that for all $a, b \in \mathfrak{R}$ with $a \leqslant b$,*

$$\mu(a, b) = \mu[a, b) = \mu(a, b] = \mu[a, b] = b - a.$$

The σ-algebra $\mathbb{A}$ is called the collection of Lebesgue measurable sets.

In other words, the Lebesgue measure of a bounded interval is its length, irrespective of which endpoints are included in the interval. You can see why Lebesgue measure is so important. It is the generalization of length in $\mathfrak{R}$ that coincides with our usual notion of the length of an interval.

1A.10. Examples and Projects. As we indicated, the Lebesgue measure of a bounded interval is its length. In this sense, then, the measure of a set is an indication of its size. As Example 1A.10B below shows, however, there are measures that are quite unrelated to size in the usual sense.

A. Let μ be Lebesgue measure for $\mathfrak{R}$. Show:

(i) For $x \in \mathfrak{R}$, $\mu(x, \infty) = \mu(-\infty, x) = \infty$.
(ii) For $x \in \mathfrak{R}$, $\mu\{x\} = 0$.
(iii) If C is a countable subset of $\mathfrak{R}$, then $\mu(C) = 0$.

B. Suppose $x \in \mathfrak{R}$. We can define a measure on the Borel sets $\mathbb{B}$ "concentrated at x" as follows: for $S \in \mathbb{B}$ define

$$\mu(S) = \begin{cases} 1 & \text{if } x \in S, \\ 0 & \text{if } x \notin S. \end{cases}$$

Show that μ is a measure on $\mathbb{B}$.

We consider next measures which may take on negative or complex values. Let X be a set and $\mathbb{A}$ be a σ-algebra in X.

1A.11. Definition. A *signed measure* μ on $\mathbb{A}$ is a function satisfying:

(i) $\mu: \mathbb{A} \to \mathfrak{R}^{\infty}$, or $\mu: \mathbb{A} \to \mathfrak{R}^{-\infty}$;
(ii) $\mu(A) < \infty$ for at least one A;
(iii) if $\mathbb{S} \subseteq \mathbb{A}$ is a pairwise disjoint, countable collection, then

$$\mu(\cup \mathbb{S}) = \sum_{S \in \mathbb{S}} \mu(S).$$

By the equality in (iii) we mean either

(a) $-\infty < \mu(\cup \mathbb{S}) < \infty$, in which case the series on the right converges for all rearrangements of its terms (hence is absolutely convergent); or
(b) $\mu(\cup \mathbb{S}) = \pm \infty$, in which case the series on the right diverges to ∞ or to $-\infty$.

The following theorem shows that every signed measure can be written as the difference of two measures. We omit its proof.

1A.12. Theorem (Jordan Decomposition Theorem). *Let X be a set and μ a signed measure on a σ-algebra $\mathbb{A}$ in X. Then there exist two measure μ^+ and μ^- on $\mathbb{A}$, at least one of which assigns finite measure to X, such that $\mu = \mu^+ - \mu^-$.*

Note that the theorem implies that μ^+ and μ^- both have domain $\mathbb{A}$ and that either $\mu^+(X)$ or $\mu^-(X)$ is finite.

Next we introduce complex-valued measures. As we shall see, they can be decomposed into ordinary measures.

1A.13. Definition. Let X be a set and $\mathbb{A}$ be a σ-algebra in X. A function $\mu: \mathbb{A} \to \mathfrak{C}$ is a *complex measure* on $\mathbb{A}$ if

(i) $\mu(A) < \infty$ for at least one A;
(ii) if $\mathbb{S} \subseteq \mathbb{A}$ is a pairwise disjoint, countable collection in $\mathbb{A}$, then $\mu(\cup \mathbb{S}) = \sum_{S \in \mathbb{S}} \mu(S)$.

Notice that implicit in condition (ii) is that $\mu(\cup \mathbb{S}) \in \mathbb{C}$, so that the series is required to converge. In particular, it is required that $\mu(X)$ is finite.

If μ is a complex measure on σ-algebra $\mathbb{A}$, then we can write $\mu = \mu_1 + i\mu_2$ for two signed measures μ_1 and μ_2 on $\mathbb{A}$ and then write μ in terms of measures on $\mathbb{A}$ as

$$\mu = \mu_1^+ - \mu_1^- + i(\mu_2^+ - \mu_2^-). \tag{1A.3}$$

We shall continue to reserve the word "measure" to refer to extended-real-valued measures, and we shall refer to the others as "complex measures."

Our final concept in this section connects functions and measures. Briefly, a real-valued function on a set X for which μ is a measure is called a "measurable function" if its *inverse* takes Borel sets to μ-measurable sets. Specifically:

1A.14. Definitions.

A. Let $f: X \to \mathfrak{R}$, and suppose μ is a measure (or complex measure) on σ-algebra $\mathbb{A}$ in X. Then f is called a *μ-measurable function* if and only if $f^{\leftarrow}[B] \in \mathbb{A}$ for every Borel set B.
B. If $f = f_1 + if_2$ is a $\mathbb{C}$-valued function on X (f_1, f_2 are real-valued), then f is called a *μ-measurable function* if both f_1 and f_2 are μ-measurable.

1A.15. Definition. Let X be a set and μ be a measure (or complex measure) on σ-algebra $\mathbb{A}$ in X. Let f and g be two μ-measurable functions on X. Then we say f equals g *μ-almost everywhere*, and we write $\underline{f = g \ \mu\text{-ae}}$, if and only if

$$\mu\{x \in X \mid f(x) \neq g(x)\} = 0.$$

IA.16. Examples and Projects.

A. Let

$$f(x) = \begin{cases} 1 & \text{if } x \in [0,1] \text{ and } x \text{ is rational,} \\ 0 & \text{otherwise.} \end{cases}$$

Show that f is equal μ-ae to the zero function. (μ is Lebesgue measure.)
B. Find a function that is equal μ-ae to the identity function on $\mathfrak{R}$, yet has the value zero at infinitely many points.

1A.17. Definition. Let X be a set and μ be a measure on σ-algebra $\mathbb{A}$ in X. Let f be a μ-measurable, real-valued function. We define for all $x \in X$,

$$\underline{f^+(x)} = \max\{f(x), 0\} \quad \text{and} \quad \underline{f^-(x)} = \max\{-f(x), 0\}.$$

1A.18. Lemma. *If X, $\mathbb{A}$, μ, and f are as in Definition* 17, *then f^+ and f^- are nonnegative μ-measurable functions and $f = f^+ - f^-$.*

PROOF. Project. □

This completes part A. In the next part we shall learn about integration with respect to different measures for $\mathfrak{R}$, and later we shall use integrals to compute expected values for experiments in quantum physics.

Part B: Integration

In this part we use measures to define integrals. The Riemann integral is usually studied in the first year of calculus, but here we will study the more general types of integrals needed to handle the discontinuous functions that arise in quantum physics.

In this part, unless otherwise noted, μ will stand for an arbitrary measure on a σ-algebra $\mathbb{A}$ for the reals $\mathfrak{R}$ with the stipulation that $\mathbb{A}$ contains the Borel sets.

1B.1. Definition. A function $f:\mathfrak{R} \to \mathfrak{R}$ is called *simple* if the image of f is finite.

Here are two examples of simple functions.

1B.2. Examples.

A. Let f be the "greatest integer function" on the interval $[0, 10]$.
B. Let f be the function

$$f(x) = \begin{cases} 0 & \text{if } x \in [0,1] \text{ and } x \text{ is rational,} \\ 1 & \text{if } x \in [0,1] \text{ and } x \text{ is irrational.} \end{cases}$$

In each of these examples it is easy to see that f has a finite image. Since we shall be interested in simple functions that are Lebesgue measurable, we point out that each f is Lebesgue measurable, because $f^{\leftarrow}\{t\}$ is a Borel set (hence a Lebesgue measurable set) for each $t \in \text{image}(f)$. In the first example, for instance, $f^{\leftarrow}\{3\} = [3, 4)$.

Now we shall define the integral of a μ-measurable simple function f. Suppose $\text{image}(f) = \{a_1, \ldots, a_n\}$. For $k = 1, \ldots, n$, let us denote $f^{\leftarrow}[a_k]$ by A_k. Then we make the following definition.

1B.3. Definition. The *μ-integral of f over a set $S \in \mathbb{A}$* is defined by

$$\int_S f \, d\mu = \sum_{k=1}^{n} a_k \mu(A_k \cap S)$$

when the sum on the right is finite.

We write $\int_S f \, d\mu = \infty$ if for some k, $\mu(A_k \cap S) = \infty$ and $a_k \neq 0$.

In the case where S is the entire domain of f, then we have $S = \bigcup_{k=1}^n A_k$, and so

$$\int_S f \, d\mu = \sum_{k=1}^{n} a_k \mu(A_k).$$

Next we define μ-integrals for functions that might not be simple. Let us consider nonnegative functions first.

1B.4. Definition. The *μ-integral* of a μ-measurable nonnegative function f over a set $S \in \mathbb{A}$ is defined by

$$\int_S f \, d\mu = \sup \left\{ \int_S h \, d\mu \,\middle|\, \begin{array}{l} h \text{ is a nonnegative, } \mu\text{-measurable, simple} \\ \text{function with } h(x) \leq f(x) \text{ for all } x \text{ in } S \end{array} \right\}.$$

We write $\int_S f \, d\mu = \infty$ if in the set on the right $\int_S h \, d\mu = \infty$ for any h or if the set on the right is unbounded.

We extend this definition to functions that may have negative image values. Recall from Definition 1A.17 the definitions of f^+ and f^- for μ-measurable function f. If f is a μ-measurable function and $S \in \mathbb{A}$, and if either $\int_S f^+ \, d\mu$ or $\int_S f^- \, d\mu$ has a finite value, then we make the following definition.

1B.5. Definition.

$$\int_S f \, d\mu = \int_S f^+ \, d\mu - \int_S f^- \, d\mu. \qquad (*)$$

We write

$$\int_S f \, d\mu = \infty \quad \text{if} \int_S f^+ \, d\mu = \infty,$$

or

$$\int_S f \, d\mu = -\infty \quad \text{if} \int_S f^- \, d\mu = \infty.$$

We say f is *μ-integrable* if and only if f is μ-measurable and both integrals on the right side of equation $(*)$ have finite values. If μ is Lebesgue measure, then $\int_S f \, d\mu$ is called the *Lebesgue integral* of f over S.

It is not difficult to connect the notion of Lebesgue integral with the notion of Riemann integral, although we will not go into the details here. Lebesgue integrals are truly a generalization of Riemann integrals in the following sense: if a function f is both Riemann and Lebesgue integrable over an interval I, then the Lebesgue and Riemann integrals of f over I are the same number. Lebesgue integration is more general in the sense that every function that is Riemann integrable over an interval I is Lebesgue integrable over I, while the converse is false. The function in Example 1B.2B is Lebesgue integrable but not Riemann integrable on $[0, 1]$.

Here we encounter a trade-off that mathematicians often face: greater generality is bought at the price of some important theorems. In this case the fundamental theorem of calculus, which relates Riemann integrals to antiderivatives, does not hold for Lebesgue integrals. For example, you may recall that to compute the Riemann integral of the function $f(x) = x^2$ over an interval $[a, b]$, it is not necessary to compute Riemann sums to approximate the integral. You can simply find an antiderivative of f, namely $F(x) = x^3/3$, and use the fundamental theorem of calculus to compute

$$\int_a^b x^2\,\mathrm{d}x = F(b) - F(a) = b^3/3 - a^3/3.$$

Of course, it usually is not so easy to find antiderivatives. In this era of computers, anyone who needs to find the value of a Riemann integral usually finds an approximation by using numerical methods to find upper and lower Riemann sums, because even the most sophisticated techniques for finding antiderivatives fail to work for most functions.

We must be careful not to confuse an *integral* such as $\int_2^3 x^2\,\mathrm{d}x$, which is a number, with an *antiderivative* such as $F(x) = x^3/3$, which is a function. It is easy to confuse them, because the fundamental theorem of calculus provides a close link between the two and because mathematicians often refer to an antiderivative as an "indefinite integral." The fundamental theorem does not apply to Lebesgue integrals, however, and when we must compute an integral in our work, we shall find it necessary to look for numerical approximations.

Integrals with respect to general measures do not satisfy all the properties that Riemann integrals do, but they do satisfy some important ones.

1B.6. Theorems. *Let μ be a measure on σ-algebra $\mathbb{A}$ for $\mathfrak{R}$ and suppose $S \in \mathbb{A}$ and f and g are functions that are μ-integrable over S.*

A. If $a \in \mathfrak{R}$, then $\int_S af\,\mathrm{d}\mu = a\int_S f\,\mathrm{d}\mu$.
B. If $f(x) \leq g(x)$ for all x in S, then

$$\int_S f\,\mathrm{d}\mu \leq \int_S g\,\mathrm{d}\mu.$$

C. For any a, b in $\mathfrak{R}$.

$$\int_S (af + bg)\,d\mu = a\int_S f\,d\mu + b\int_S g\,d\mu.$$

D. If T and U are disjoint μ-measurable sets in S, then

$$\int_{T\cup U} f\,d\mu = \int_T f\,d\mu + \int_U f\,d\mu.$$

PROOF. The proof of part B is in the Coaching Manual; the proofs of the other parts are omitted. □

1B.7. Theorem (The Monotone Convergence Theorem). *Let μ be a measure on σ-algebra* $\mathbb{A}$ *for* $\mathfrak{R}$ *and suppose* $S\in\mathbb{A}$. *If* $\langle\!\langle f_n \rangle\!\rangle$ *is a sequence of μ-integrable functions converging pointwise to f on S, and for each* $n\in\mathbb{N}$, $0\leqq f_n(x)\leqq f(x)$ *for all* $x\in S$, *then*

$$\lim_n \left\langle\!\!\left\langle \int_S f_n\,d\mu \right\rangle\!\!\right\rangle = \int_S f\,d\mu.$$

PROOF. Omitted. □

1B.8. Theorems. *Let μ be a measure on the Borel sets and S a Borel set.*

A. Suppose f is a function μ-integrable over S, and $f(x)=0$ *μ-ae on S. Then*

$$\int_S f\,d\mu = 0.$$

[*Recall that* $f(x)=0$ *μ-ae on S if and only if* $\mu\{x \mid x\in S \text{ and } f(x)\neq 0\}=0$.]

B. The following is a partial converse of (A) above. If f is a nonnegative μ-measurable function and $\int_S f\,d\mu = 0$, *then* $f(x)=0$ *μ-ae on S.*

PROOF. Project. □

1B.9. Examples and Projects. In these examples μ is Lebesgue measure and I is the interval [0.1].

A. Consider the function

$$f_0(x) = \begin{cases} 1 & \text{if } x \text{ is rational and } x\in I,\\ 0 & \text{if } x \text{ is irrational and } x\in I.\end{cases}$$

The function f_0 is not Riemann integrable over I. Show, however, that f_0 is Lebesgue integrable over I, and compute $\int_I f_0\,d\mu$.

B. Consider the function

$$f_1(x) = \begin{cases} x & \text{if } x \text{ is rational and } x\in I,\\ 0 & \text{if } x \text{ is irrational and } x\in I.\end{cases}$$

Use the result of Theorem 1B.8A to compute the Lebesgue integral

$$\int_I f_1 \, d\mu.$$

C. Let $S = [0, \pi]$. Consider the function

$$f_2(x) = \begin{cases} 1/q & \text{if } x = p/q \text{ in lowest terms } (p, q \in \mathbb{Z}) \text{ and } x \in S, \\ 2 & \text{if } x \text{ is irrational and } x \in S. \end{cases}$$

Use the result in Theorem 1B.8A to compute the Lebesgue integral $\int_S f_2 \, d\mu$.

Now let us return to the firefly that we put into a box at the beginning of this chapter. Recall our experiment $E = \{0, 1, \ldots, 20\}$. Let ω be the weight on E given in Project 1A.3, and let μ be the measure for $\mathfrak{R}$ defined by equation (1A.1). We then have the following result.

1B.10. Lemma.

$$\operatorname{Exp}(E, \omega) = \sum_{x \in E} x\omega(x) = \int_{\mathfrak{R}} I_{\mathfrak{R}} \, d\mu,$$

where $I_{\mathfrak{R}}(x) = x$ is the identity function on $\mathfrak{R}$.

PROOF. Project. □

This connection between expected values and the integral of the identity function is one of the cornerstones of the orthodox formulation of quantum mechanics. We will return to this theme in Chapters 6, 7, and 8. The main idea will be to consider measures for $\mathfrak{R}$ that depend on the eigenvalues of certain linear operators. If the eigenvalues form a discrete set, then a variable that can assume only these eigenvalues is called "quantized." Our measure will assign to each Borel set a value that depends on how many eigenvalues are in the set. We then show how we can formulate the notion that if we physically measure a quantized variable, such as energy, the probability of obtaining a value in a given interval I depends in part on the measure of I, that is, on how many eigenvalues are in I. The expected value of the variable will then be computed using a Lebesgue integral of the identity function with respect to this eigenvalue-dependent measure.

To conclude this chapter we define the integral of a complex-valued function with respect to a complex measure.

Let μ be a complex measure on σ-algebra $\mathbb{A}$ for $\mathfrak{R}$ and suppose f is a real-valued, μ-measurable function on $\mathfrak{R}$. Suppose further that μ is written in terms of (real) measures, as in equation (1A.3): $\mu = \mu_1^+ - \mu_1^- + i(\mu_2^+ - \mu_2^-)$. Suppose also that $S \in \mathbb{A}$, that either $\int_S f \, d\mu_1^+$ or $\int_S f \, d\mu_1^-$ is finite, and that either $\int_S f \, d\mu_2^+$ or $\int_S f \, d\mu_2^-$ is finite. Then we make the following definition.

1B.11 Definition. The *μ-integral* of f over set $S \in \mathbb{A}$ is

$$\underline{\int_S f\,\mathrm{d}\mu} = \int_S f\,\mathrm{d}\mu_1^+ - \int_S f\,\mathrm{d}\mu_1^- + i\left(\int_S f\,\mathrm{d}\mu_2^+ - \int_S f\,\mathrm{d}\mu_2^-\right)..$$

We write $\int_S f\,\mathrm{d}\mu = \infty$ if any one of the four integrals on the right is infinite.

Finally, if μ is a complex measure, and $f = f_1 + if_2$ is a μ-measurable complex-valued function (f_1 and f_2 are real-valued), and $S \in \mathbb{A}$, we make the following definition.

1B.12. Definition.

$$\underline{\int_S f\,\mathrm{d}\mu} = \int_S f_1\,\mathrm{d}\mu + i\int_S f_2\,\mathrm{d}\mu.$$

Notice that if μ is a complex measure and f is a μ-measurable, *bounded* function, then

$$\left|\int_S f\,\mathrm{d}\mu\right| \leqq \left|\int_S f_1\,\mathrm{d}\mu\right| + \left|\int_S f_2\,\mathrm{d}\mu\right| \leqq \sup\{|f_1(x)| \mid x \in \mathfrak{R}\}|\mu(\mathfrak{R})|$$

$$+ \sup\{|f_2(x)| \mid x \in \mathfrak{R}\}|\mu(\mathfrak{R})| < \infty,$$

since, as we remarked after Definition 1A.13, $|\mu(\mathfrak{R})| < \infty$.

Finally, we state without proof a theorem we shall need in Chapter 6A.

1B.13. Theorem. *If μ_1 and μ_2 are finite measures on σ-algebra $\mathbb{A}$ in $\mathfrak{R}$, and $t_1, t_2 \in \mathfrak{R}$, then $t_1\mu_1, t_2\mu_2$, and $t_1\mu_1 + t_2\mu_2$ are all signed measures on $\mathbb{A}$. Further, if $f : \mathfrak{R} \to \mathfrak{R}$ is integrable over $\mathfrak{R}$ with respect to all of these signed measures, then*

$$\int_{\mathfrak{R}} f\,\mathrm{d}(t_1\mu_1 + t_2\mu_2) = t_1\int_{\mathfrak{R}} f\,\mathrm{d}\mu_1 + t_2\int_{\mathfrak{R}} f\,\mathrm{d}\mu_2.$$

This completes our introduction to measures and integrals. We turn now to Hilbert spaces, and we will consider the deep connection between Hilbert spaces and measures in Chapter 7 when we discuss the Spectral Theorem. In the meantime we will have occasion to use integrals to see some important examples of Hilbert spaces.

CHAPTER 2

Hilbert Space Basics

We assume that you know the definition of a vector space $\mathscr{V} = (V, +, \cdot)$ over the field $\mathfrak{C}$ of complex numbers. Perhaps you recall defining an "inner product" between vectors in a vector space and using the inner product to consider angles between vectors. A "Hilbert space" is a vector space with an inner product. We begin by defining an inner product space.

2.1. Definition. A structure $\mathscr{V} = (V, +, \cdot \langle , \rangle)$ is an *inner product space* if

(i) $(V, +, \cdot)$ is a vector space (over the field of complex numbers);
(ii) $\langle , \rangle$ is a function that associates a complex number with every pair of vectors in V subject to the following rules: for all $x, y, z \in V$ and $\lambda \in \mathfrak{C}$,

(a) $\langle x, y \rangle = \langle y, x \rangle^*$ ($*$ is complex conjugate);
(b) $\langle x + y, z \rangle = \langle x, z \rangle + \langle y, z \rangle$ and $\langle x, y + z \rangle = \langle x, y \rangle + \langle x, z \rangle$;
(c) $\langle \lambda x, y \rangle = \lambda \langle x, y \rangle$ (we follow the usual practice of writing λx for $\lambda \cdot x$);
(d) $\langle x, x \rangle$ is a nonnegative real number, and $\langle x, x \rangle = 0$ only if x is the zero vector in V.

We call $\langle x, y \rangle$ the *inner product* of x and y.

We will adopt the usual practice of denoting an inner product space simply by its underlying set V, since there will seldom be any doubt about which inner product structure is under discussion.

2.2. Lemma. *If V is an inner product space and $x, y \in V$ and $\lambda \in \mathfrak{C}$, then* $\langle x, \lambda y \rangle = \lambda^* \langle x, y \rangle$.

PROOF. Project. □

2.3. Examples and Projects.

A. Suppose n is a positive integer. Let $V = \mathfrak{C}^n$ be the set of complex n-tuples organized into a vector space under pointwise addition and scalar multiplication. For $x = (x_1, \ldots, x_n)$ and $y = (y_1, \ldots, y_n)$ in V, define $\langle x, y \rangle = \sum_{k=1}^{n} x_k y_k^*$. Show that $\langle\,,\rangle$ is an inner product. We call this inner product space *complex n-space* and denote it by $\mathfrak{C}^n$.

B. Let V be defined by

$$V = \{\langle\!\langle x_k \rangle\!\rangle \mid \langle\!\langle x_k \rangle\!\rangle \text{ is a sequence of complex numbers, and the series } \sum_{k=1}^{\infty} |x_k|^2 \text{ converges in } \mathfrak{R}\}.$$

Using standard results for convergent series, we can show that V can be organized into a vector space under pointwise addition and scalar multiplication. Define for each $x, y \in V$

$$\langle x, y \rangle = \sum_{k=1}^{\infty} x_k y_k^*.$$

Show that the series converges. (The calculation is a bit tricky.) Then show that $\langle\,,\rangle$ is an inner product. (This is more straightforward.) This inner product space is denoted by l^2. (Some authors write l_2.)

C. Suppose μ is a complex Borel measure for $\mathfrak{R}$. Suppose also that $S = (a, b)$ is a nonempty interval in $\mathfrak{R}$. Define

$$V = \{\psi \mid \psi : S \to \mathfrak{C} \text{ and } |\psi|^2 \text{ is } \mu\text{-integrable over } S\}.$$

We can define pointwise addition and scalar multiplication on V by

(i) $(\psi_1 + \psi_2)(t) = \psi_1(t) + \psi_2(t)$ for all $t \in S$;
(ii) $(\lambda \cdot \psi)(t) = \lambda \psi(t)$ for all $t \in S$, $\lambda \in \mathfrak{C}$.

It can be shown that $(V, +, \cdot)$ is a vector space. For all $\psi_1, \psi_2 \in V$, define

$$\langle \psi_1, \psi_2 \rangle = \int_S \psi_1 \psi_2^* \, d\mu.$$

With a proof similar to the one in Example B above it can be shown that $\int_S |\psi_1 \psi_2^*| \, d\mu \leqq \frac{1}{2}(\int_S |\psi_1|^2 \, d\mu + \int_S |\psi_2|^2 \, d\mu)$, so that $\langle \psi_1, \psi_2 \rangle$ is a complex number.

A slight complication arises, however, if we try to establish property (ii)(d) in the Definition 2.1 of an inner product. That property states that $\langle \psi, \psi \rangle = 0$ implies $\psi = 0$. By Theorem 1B.8B, however, we know only that if $\langle \psi, \psi \rangle = \int_S \psi \psi^* \, d\mu = \int_S |\psi|^2 \, d\mu = 0$, then $|\psi|^2 = 0$ μ-almost everywhere on S. The way to get around this difficulty is to define an equivalence relation on V by

$$\psi_1 \sim \psi_2 \text{ if and only if } |\psi_1 - \psi_2| = 0 \ \mu\text{-ae on } S.$$

Denote by $\hat{V}$ the set of equivalence classes determined by members of V.

We leave as a project the varification that the pointwise addition and scalar multiplication and the inner product defined on V can be transferred in a natural way to $\hat{V}$. The resulting inner product space is denoted in many books by $\mathscr{L}_2(a,b)$ or $\mathscr{L}^2(a,b)$ when μ is Lebesgue measure.

We can use an inner product to define a length in a vector space. For the rest of this chapter we shall let V be an arbitrary inner product space unless otherwise noted.

2.4. Definition. The *inner product norm* on V is a function from V to $\mathfrak{R}$ defined for every $x \in V$ by $\|x\| = \sqrt{\langle x, x \rangle}$.

2.5. Project. Suppose $n \in \mathbb{N}$. Show that for $x = (x_1, \ldots, x_n) \in \mathfrak{C}^n$, $\|x\|^2 = \sum_{k=1}^n |x_k|^2$. This is why $\|x\|$ is called the "length" of x as well as the *norm of* x.

2.6. Theorem. *The inner product norm satisfies the following for all $x, y \in V$ and $\lambda \in \mathfrak{C}$:*

(i) $\|x\| \geqq 0$, *and equality holds only if* $x = 0$;
(ii) $\|\lambda x\| = |\lambda| \, \|x\|$;
(iii) $\|x + y\|^2 + \|x - y\|^2 = 2\|x\|^2 + 2\|y\|^2$.

The equation in (iii) is called the parallelogram law. If $V = \mathfrak{C}^2$ we can illustrate this law (see Figure 2.1), which states that the sum of the squares of the diagonals of a parallelogram equals the sum of the squares of its sides.

PROOF. Project. □

Inner products are also used to define orthogonality (also called perpendicularity) in vector spaces.

2.7. Definition. Two vectors $x, y \in V$ in inner product space V are said to be *orthogonal* to each other [written $x \perp y$] if $\langle x, y \rangle = 0$.

If $S, T \subseteq V$, we say *S is orthogonal to T* [written $S \perp T$] if for all $x \in S$ and $y \in T$, $x \perp y$.

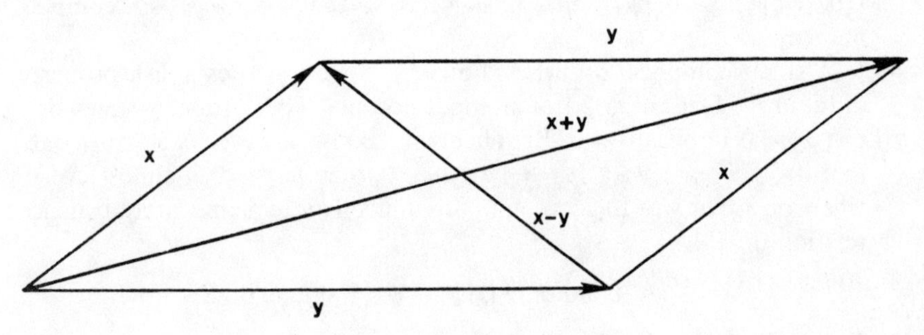

Figure 2.1. The parallelogram law.

2.8. Lemma. *For all* $x \in V$, $x \perp 0$.

PROOF. Project. □

2.9. Lemma.

$$\text{If } x \perp y, \text{ then } \|x+y\|^2 = \|x\|^2 + \|y\|^2.$$

(*A sketch using* $\mathfrak{C}^2$ *should make it clear to you why this is called the Pythagorean theorem.*)

PROOF. A straightforward computation using $\langle x, y \rangle = 0$. □

2.10. Definition. A set $S \subseteq V$ is called *orthonormal* if its members are pairwise orthogonal and all have norm one.

2.11. Theorem (Bessel's Inequality). *If* $\{x_1, \ldots, x_p\}$ *is an orthonormal set in* V *then for all* $y \in V$,

$$\sum_{k=1}^{p} |\langle x_k, y \rangle|^2 \leqq \|y\|^2.$$

PROOF. Project.** □

2.12. Corollaries to Bessel's Inequality.

A. *If* $\langle\langle x_k \rangle\rangle$ *is an orthonormal sequence in* V (*i.e.*, $\{x_k \mid k \in \mathbb{N}\}$ *is an orthonormal set*), *then for all* $y \in V$,

$$\sum_{k=1}^{\infty} |\langle x_k, y \rangle|^2 \leqq \|y\|^2.$$

B. *If* $x \in V$ *and* $\|x\| = 1$, *then for all* $y \in V$, $|\langle x, y \rangle| \leqq \|y\|$.

PROOF. Project. □

2.13. Theorem (The Cauchy–Schwarz Inequality). *For all* $x, y \in V$, $|\langle x, y \rangle| \leqq \|x\| \, \|y\|$.

PROOF. Project. □

2.14. Theorems. *For all* $x, y \in V$,

A. $\|x+y\| \leqq \|x\| + \|y\|$, *and*
B. $|\,\|x\| - \|y\|\,| \leqq \|x-y\|$.

Property (*A*) *is called the triangle inequality.*

PROOF. Project. □

2.15. Definitions.

A. A sequence $\langle\langle x_k \rangle\rangle$ in V *converges in norm* to a vector $y \in V$ if and only if $\lim_{n\to\infty} \|x_n - y\| = 0$.
B. A sequence $\langle\langle x_k \rangle\rangle$ in V is *Cauchy* if and only if for every $\varepsilon > 0$, there is an $N_\varepsilon \in \mathbb{N}$ such that for all $n, m > N_\varepsilon$, $\|x_n - x_m\| < \varepsilon$.

2.16. Theorem. *Every sequence in V that converges in norm to a vector in V is Cauchy.*

PROOF. Project. □

The converse to Theorem 2.16 is not true for general inner product spaces. We single out those spaces for which it is true.

2.17. Definition. An inner product space V is *complete* if and only if every Cauchy sequence in V converges in norm to some vector in V.

2.18. Definition. A *Hilbert space* is a complete inner product space.

For the rest of this chapter H will stand for an arbitrary Hilbert space unless otherwise noted.

2.19. Examples and Projects. The examples in 2.3 are all Hilbert spaces. The proofs of completeness in Examples 2.3A and 2.3B are projects. (The proof for 2.3B is at level $*$.) The proof of completeness in Example 2.3C is rather involved and is omitted.

Recall that finite dimensional vector spaces all have finite bases. That is, if V is a finite dimensional vector space, there is a finite linearly independent set $B \subseteq V$ such that every vector in V can be written uniquely as a linear combination of the members of B. A most important fact about a general Hilbert space is that, as a vector space, it need not have a finite basis. We turn to that issue now, and we begin by considering infinite sums of vectors.

2.20. Definition. A sequence $\langle\langle x_k \rangle\rangle$ in H is called *summable* if and only if there exists $x \in H$ such that the sequence $\langle\langle \sum_{k=1}^{n} x_k \rangle\rangle$ converges in norm to x. In that case we write $x = \sum_{k=1}^{\infty} x_k$.

2.21. Theorems.

A. *If $\langle\langle x_k \rangle\rangle$ is a sequence in H that converges in norm to $x \in H$, then $\lim_{k\to\infty} \|x_k\| = \|x\|$.*
B. *If $\langle\langle x_k \rangle\rangle$ and $\langle\langle y_k \rangle\rangle$ are sequences in H that converge in norm to x and y, respectively, then*

(i) $\lim_{k\to\infty} \langle x_k, y_k \rangle = \langle x, y \rangle$ (*we have written* $\langle x_k, y_k \rangle$ *instead of* $\langle\!\langle \langle x_k, y_k \rangle \rangle\!\rangle$);
(ii) $\lim_{k\to\infty} \langle\!\langle x_k + y_k \rangle\!\rangle = x + y$;
(iii) *for all* $\lambda \in \mathfrak{C}$, $\lim_{k\to\infty} \langle\!\langle \lambda x_k \rangle\!\rangle = \lambda x$.

C. *If* $\langle\!\langle x_k \rangle\!\rangle$ *is a summable sequence in* H *and if* $y \in H$, *then*

$$\left\langle \sum_{k=1}^{\infty} x_k, y \right\rangle = \sum_{k=1}^{\infty} \langle x_k, y \rangle.$$

D. *If* $S = \{x_1, x_2, \ldots\}$ *is a countable orthonormal set in* H *and* $x = \sum_{k=1}^{\infty} \lambda_k x_k$, *then*

(i) *for all* $k \in \mathbb{N}$, $\lambda_k = \langle x, x_k \rangle$, *and*
(ii) $\|x\|^2 = \sum_{k=1}^{\infty} |\lambda_k|^2$.

E. *If* $S = \{x_1, x_2, \ldots\}$ *is a countable set of nonzero vectors in* H *and* $\sum_{k=1}^{\infty} \|x_k\| < \infty$, *then there exists* $x \in H$ *such that* $x = \sum_{k=1}^{\infty} x_k$. *If* S *is pairwise orthogonal and* $x = \sum_{k=1}^{\infty} x_k$, *then* $\|x\|^2 = \sum_{k=1}^{\infty} \|x_k\|^2$.

PROOF. Project. □

Theorems 2.21D and E are a generalization of the Pythagorean theorem.

2.22. Definitions.

A. A countable subset $S = \{x_1, \ldots\}$ of H is called *linearly independent* if and only if for all sequences $\langle\!\langle \lambda_k \rangle\!\rangle$ in $\mathfrak{C}$, $\sum_{k=1}^{\infty} \lambda_k x_k = 0$ implies $\lambda_k = 0$ for all $k \in \mathbb{N}$.
B. A *basis* for Hilbert space H is a maximal orthonormal subset of H.

2.23. Theorem. *Every finite or countably infinite orthonormal set in* H *is linearly independent in* H.

PROOF. Project. □

There is a powerful result known as Zorn's lemma that can be used to show that every Hilbert space has a basis, possibly one of high cardinality. In this book we shall assume that all of our Hilbert space bases are finite or *countably* infinite. Such Hilbert spaces are usually described as *separable.*

2.24. Theorem. *Suppose* H *is a Hilbert space and* B *is a countable, orthonormal subset of* H. *Then the following are equivalent:*

(i) B *is a basis for* H;
(ii) *if* $x \in H$, *then* $x \perp b$ *for all* $b \in B$ *if and only if* $x = 0$;
(iii) *if* $x \in H$, *then* $x = \sum_{b\in B} \langle x, b \rangle b$;
(iv) *if* $x, y \in H$, *then* $\langle x, y \rangle = \sum_{b\in B} \langle x, b \rangle \langle b, y \rangle$;
(v) *if* $x \in H$, *then* $\|x\|^2 = \sum_{b\in B} |\langle x, b \rangle|^2$.

The sum in (iii) is called the Fourier expansion of x with respect to basis B. The equalities in (iv) and (v) are both referred to as Parseval's identity.

PROOF. Project.** □

Recall that for finite dimensional vector spaces it is common to define a basis as a *linearly independent spanning set*. The following theorem shows that for finite dimensional Hilbert spaces maximal orthonormal sets are linearly independent spanning sets.

2.25. Theorem. *If B is a finite basis for Hilbert space H, then B is a basis (linearly independent spanning set) for vector space H.*

PROOF. We know from Theorem 2.23 that B is linearly independent. That it spans H follows from Theorem 2.24(iii). □

2.26. Theorem. *All bases for a given Hilbert space have the same cardinality.*

PROOF. The proof of this theorem involves a considerable amount of mathematics, and not less so because we are requiring that all our bases be countable. To go into the details required would take us too far astray from the main purpose of this book, so we omit the proof.

2.27. Definition. The *dimension* of a Hilbert space is the cardinality of any one (hence all) of its bases.

2.28. Examples and Projects. Show that the Hilbert space of Example 2.3B is infinite dimensional. Show also that the Hilbert space of Example 2.3C is infinite dimensional if μ is Lebesgue measure and $S = (0, 1)$.

Recall that for $n \in \mathbb{N}$ the Hilbert space $\mathfrak{C}^n$ has basis $B = \{b_1, \ldots, b_n\}$, where $b_k = (0, 0, \ldots, 1, \ldots, 0)$ with 1 in the kth coordinate. We shall call this the *standard basis* for $\mathfrak{C}^n$.

A word of caution is in order. If H is a Hilbert space of dimension $n \in \mathbb{N}$, and if B is an arbitrary orthonormal set of cardinality n, then B is a basis for H; in other words, it is a maximal orthogonal set. The analogous statement for infinite dimensional spaces is false. That is, a countably infinite orthonormal subset of an infinite dimensional Hilbert space is *not necessarily maximal.* You can convince yourself of this merely by considering any infinite proper subset of a basis for an infinite dimensional space.

While Hilbert spaces would be of very little value in physics if it were not for the infinite dimensional ones, there are some important ideas in the foundations of physics that can be discussed using only finite dimensional Hilbert spaces. In the next chapter we reveal the heart of Heisenberg's uncertainty principle using a Hilbert space of dimension two.

CHAPTER 3

The Logic of Nonclassical Physics

Introduction

In this chapter we introduce a mathematical formulation for the foundations of quantum physics. Our formulation incorporates three key ideas:

(1) Physical variables come in two varieties, those with a continuous range of possible values and those with a discrete range of possible values.
(2) There is an irreducible probabilism inherent in nature.
(3) There are pairs of physical variables that cannot be simultaneously measured to arbitrary degrees of accuracy.

It should be emphasized that historically quantum physics was not born of this or any other neatly stated set of assumptions. Rather it evolved from a potpourri of mathematical formulas resembling classical laws and imaginative recipes for using those formulas. It is only after many years of studying the applications of the rules for quantum physics that theoreticians are beginning to extract the foundational assumptions on which these rules rest.

Part A: Manuals of Experiments and Weights

A "physical system" is anything on which we perform experiments. Examples of physical systems are a pendulum, a little black box with two rods sticking out of it, or even the entire universe. For example, at the beginning of Chapter 1 we considered a physical system consisting of a firefly in a box. Now we consider a slightly more sophisticated system.

Consider a box with a clear plastic window at the front and another one

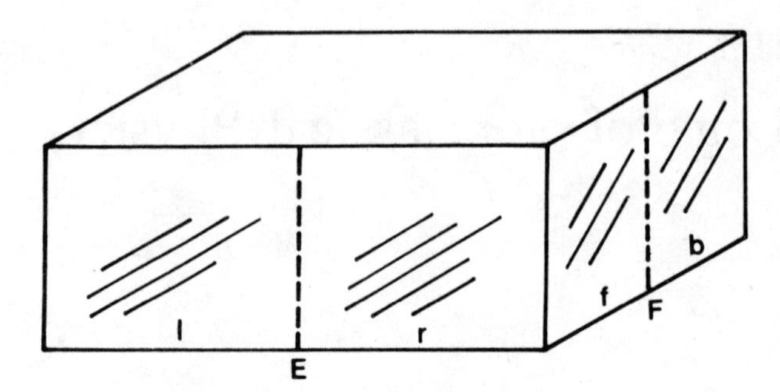

Figure 3A.1 A physical system consisting of a box with two marked windows and a firefly inside.

on one side. See Figure 3A.1. Suppose each window has a thin vertical line drawn down the center to divide the window in half. Place a firefly in the box. This is our physical system.

We shall consider two experiments on the system. Experiment E is: Look at the front window. The outcomes of E will be:

l = see a light in the left half of the window,

r = see a light in the right half of the window,

n = see no light.

Let us denote this experiment by $E = \{l, r, n\}$. A second experiment F is: Look at the side window. The outcomes of F will be:

f = see a light in the left half of window (near front of box),

b = see a light in the right half of window (near back of box),

n = see no light.

Let us denote this experiment by $F = \{f, b, n\}$. In Figure 3A.2 we show a diagram that illustrates our two experiments and their outcomes. After we consider a few definitions we shall see why our diagram was drawn in this particular way.

Our approach to quantum logics is based on the notion of a *manual*, which is a mathematical representation of a book of laboratory experiments. Before

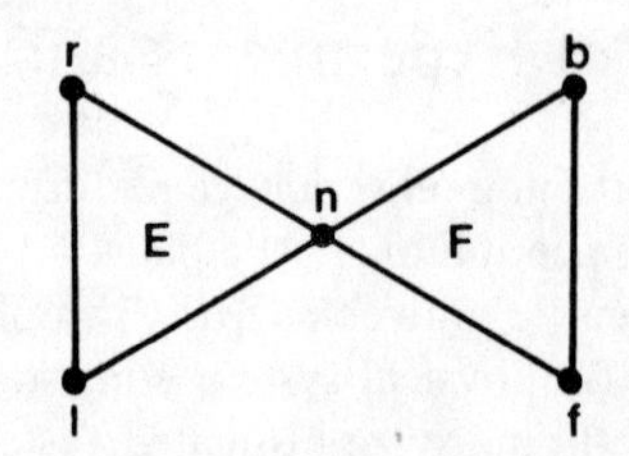

Figure 3A.2. The "bowtie" manual.

defining a manual, however, we start with a more primitive notion, which we call a *quasimanual.*

3A.1. Definitions.

A. A *quasimanual* $\mathscr{Q}$ is a nonempty collection of nonempty sets called *experiments.* The members of the experiments are called *outcomes.* The set of all outcomes is denoted by $X_{\mathscr{Q}}$.
B. An *event* in quasimanual $\mathscr{Q}$ is a subset of an experiment in $\mathscr{Q}$.

We say we *test for event A* by performing an experiment that contains A. If we test for A and obtain an outcome in A, we say *event A occurred.*

3A.2. Definitions. Suppose $\mathscr{Q}$ is a quasimanual.

A. Two events A, B in $\mathscr{Q}$ are said to be *orthogonal*, denoted $A \perp B$, if they are disjoint subsets of a single experiment in $\mathscr{Q}$. (For outcomes x and y of $\mathscr{Q}$ we write $x \perp y$ to mean $\{x\} \perp \{y\}$.)
B. If A, B are orthogonal events in $\mathscr{Q}$ and $A \cup B$ is an experiment in $\mathscr{Q}$, then we say that A and B are *orthogonal complements* in $\mathscr{Q}$. We denote this by A oc B.

3A.3. Definition. A *manual* is a quasimanual $\mathfrak{M}$ which satisfies the following:

(i) If A, B, C, D are events in $\mathfrak{M}$ with A oc B, B oc C, and C oc D, then $A \perp D$.
(ii) If $E, F \in \mathfrak{M}$ and $E \subseteq F$, then $E = F$.
(iii) If x, y, z are outcomes in $\mathfrak{M}$ with $x \perp y$, $y \perp z$, $z \perp x$, then $\{x, y, z\}$ is an event in $\mathfrak{M}$.

Property (ii) ensures that experiments are maximal events. Property (iii) is called the "orthocoherence property." (In some literature "manuals" are defined without the orthocoherence requirement.)

We shall discuss the rationale for property (i) by referring to Figure 3A.3.

Suppose we test for event A by performing experiment E, and A occurs. Then we know that B did not occur. Thus, if we had performed experiment F, then C would have occurred; so, if we had performed experiment G, event D could *not* have occurred. In summary, if we test for A, and A occurs, then testing for D would result in D not occurring. A similar reasoning shows that if D occurs when tested, then A cannot occur when tested at the same time. Hence A and D are events that bear a special relationship to each other through E, F, and G, and it is not unnatural to require that there is a single experiment H that contains both A and D, so that $A \perp D$ in $\mathfrak{M}$.

The word "manual" was chosen to reflect our attitude that our knowledge about a physical system depends entirely on the set of all experiments that can be performed on the system and the relationships between those experiments. Thus, a laboratory manual containing all the details for conducting all known experiments is the basic tool for learning about the

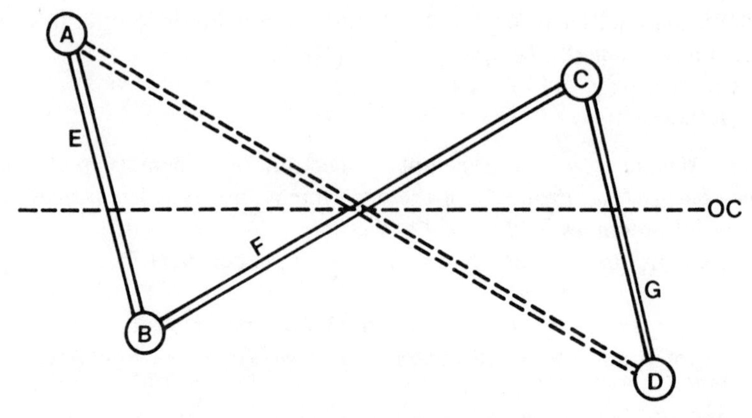

Figure 3A.3. An oc diagram. Solid double lines connect events that are orthocomplements. In a manual if A, B, C and D are orthocomplements as indicated, then A must be orthogonal to D.

physical universe. While much of what follows can be proved for quasimanuals, we shall deal only with manuals. The additional structure we are requiring will be of great convenience in our mathematical development and will not unduly restrict our choice of examples.

3A.4. Examples and Project.

A. Let X be a set with at least two members, and let $\mathcal{Q}$ be the collection of all nonempty subsets of X. Show that $\mathcal{Q}$ is a quasimanual but not a manual.

B. Let $\mathfrak{R}$ be the set of real numbers and let $\mathfrak{M}$ be the collection of all countable partitions of $\mathfrak{R}$ into Borel sets. Show that $\mathfrak{M}$ is a manual. We call it the *partition manual on* $\mathfrak{R}$.

C. Let H be a finite dimensional Hilbert space and let $\mathscr{F}(H)$ be the collection of orthonormal bases for H. Show that $\mathscr{F}(H)$ is a manual. We call it the *frame manual for H*. Notice that two outcomes in a frame manual $\mathscr{F}(H)$ are orthogonal in the manual if and only if they are orthogonal unit vectors in the Hilbert space H. Thus, although we use the word orthogonal in two different ways, in a frame manual the meanings coincide.

D. In the preceding example we required that H be finite dimensional because the argument for an infinite dimensional space requires concepts that we shall not see until Chapter 4. It is important to note, however, that the collection $\mathscr{F}(H)$ of orthonormal bases for any separable Hilbert space, even one of infinite dimension, forms a manual, which we call the *frame manual* for H.

If we have a manual of experiments for a physical system, and a new experiment is made up, we must fit it into the manual in an appropriate way. We define that appropriate way as follows.

3A.5. Definition. Manual $\mathfrak{M}_2$ is a *refinement* of manual $\mathfrak{M}_1$ if there is an injection $\varphi: X_{\mathfrak{M}_1} \to X_{\mathfrak{M}_2}$ such that for every experiment $E \in \mathfrak{M}_1$, $\varphi(E) = \{\varphi(x) | x \in E\}$ is an event in $\mathfrak{M}_2$. We call φ a *refinement morphism* from $\mathfrak{M}_1$ to $\mathfrak{M}_2$. We write $\mathfrak{M}_1 < \varphi\, \mathfrak{M}_2$ to denote that $\mathfrak{M}_2$ is a refinement of $\mathfrak{M}_1$ under refinement morphism φ.

Thus, if we design new experiments for a physical system, we must refine our laboratory manual by making sure that the old experiments are at least events in the new manual.

3A.6. Definition. If $\mathfrak{M}$ is a manual, A, B, and C are events in $\mathfrak{M}$, and $A \text{ oc } B$ and $B \text{ oc } C$, then we say that A and C are *operationally perspective*, which we denote by $A\ op\ C$.

3A.7. Project. Suppose A and C are operationally perspective in manual $\mathfrak{M}$. Show that A occurs (respectively, does not occur) when tested precisely at those instants when C occurs (respectively, does not occur) when tested.

Now we turn back to Figure 3A.2. This sketch is an orthogonality diagram, which shows the two experiments E and F with orthogonal pairs of outcomes connected by line segments. Notice that l and f are not connected by a line segment, because there is no experiment that contains them both. By the way, it is this diagram that suggests why we refer to this manual as the *bow tie manual.*

Next we consider the notion of "compatibility" of experiments.

3A.8. Definitions.

A. A collection $\mathbb{E}$ of events in a manual $\mathfrak{M}$ is *compatible* if $\cup \mathbb{E}$ is an event in $\mathfrak{M}$.
B. A collection O of outcomes is *compatible* if O is an event.

Thus, a collection of events in a manual is compatible if and only if there is one experiment that contains all those events. Performing that one experiment tests all the events simultaneously.

3A.9. Definition. A manual is called *classical* if every pair of events is compatible. Equivalently, a manual is classical if every collection of outcomes is an event.

It was the classical assumption (perhaps unspoken) prior to this century that any pair of experiments on a physical system can be performed simultaneously, at least in theory. Quantum physics challenged that assumption, and in our simple firefly experiment we see the heart of the simultaneity issue. If we believe that the bow tie manual is the best possible characterization of the firefly system, then we believe our firefly in a box is a nonclassical physical system, because E and F cannot be performed simultaneously. If we

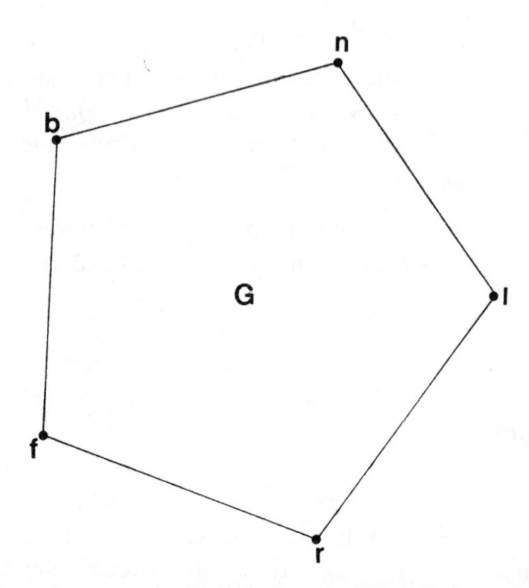

Figure 3A.4. A classical manual for the firefly-in-a-box system.

believe that we can perform E and F simultaneously, perhaps by posting two observers, one at each window, and assuming that they can communicate instantaneously, then we believe our system is classical and can be characterized by a classical manual. In other words, if we have a nonclassical manual $\mathfrak{M}_1$ describing a system that we believe is classical, then we believe that there exists a classical manual $\mathfrak{M}_2$ with $\mathfrak{M}_1 <\varphi\, \mathfrak{M}_2$, so that for all $E, F \in \mathfrak{M}_1$, $\varphi(E)$ and $\varphi(F)$ are compatible in $\mathfrak{M}_2$.

In Figure 3A.4 we have drawn an orthogonality diagram for a classical manual characterizing the firefly system. Let us examine more closely this idea of a manual "characterizing" a physical system. A manual is a set of experiments. To completely characterize a physical system, it does not seem sufficient merely to show the experiments that can be performed on it. We should also provide some information about how the system is likely to behave when we examine it with the experiments. In other words, to characterize a physical system, we should provide not only a manual but also information about the likely occurrence or nonoccurrence of the events in the manual. This brings us to the idea of the "states" of a physical system.

In what follows we need the notion of an ordered sum.

3A.10. Definitions. Suppose E is a set and ω is a function from E to the nonnegative real numbers. We define

$$\underline{\sum_{x \in E} \omega(x)} = \text{lub}\left\{ \sum_{x \in S} \omega(x) \,|\, S \text{ is a finite subset of } E \right\}.$$

We write $\sum_{x \in E} \omega(x) = \infty$ if the set on the right is unbounded. If $A \subseteq E$, we abuse notation by writing $\underline{\omega(A)}$ for $\sum_{x \in A} \omega(x)$.

3A.11. Definitions.

A. A *weight* on a quasimanual $\mathcal{Q}$ is a function $\omega: X_{\mathcal{Q}} \to [0, 1]$ such that for every experiment E in $\mathcal{Q}$, $\omega(E) = 1$.
B. The collection of all weights on a quasimanual $\mathcal{Q}$ is denoted by $\underline{\Omega_{\mathcal{Q}}}$.

We shall be concerned mainly with weights on manuals. We shall consider that a physical system characterized by a manual $\mathfrak{M}$ is associated at every instant in time with a weight function ω, so that if we test for event A at that time by performing an experiment that contains A, then the *probability* that A will occur is $\omega(A)$. Naturally, if we perform experiment E, then no matter what state the system is in, we shall certainly obtain an outcome in E That is why we require that $\omega(E) = 1$.

3A.12. Example and Project. A weight ω on the bow tie manual is determined by its values on r, n, and b. That is because $\omega(l) = 1 - (\omega(n) + \omega(r))$ and $\omega(f) = 1 - (\omega(n) + \omega(b))$. Draw a three-dimensional coordinate system with axes α, β, γ, and find the region of $\mathfrak{R}^3$ that represents

$$w_{\mathfrak{M}} = \{(\alpha, \beta, \gamma) \mid \omega \in \Omega_{\mathfrak{M}} \text{ and } \alpha = \omega(r), \quad \beta = \omega(b), \text{ and } \gamma = \omega(n)\}.$$

[The figure will be most easily recognized if you draw the α-β-γ axis system in the perspective: α comes "out of the paper," β is horizontal, and γ is vertical.]

We turn next to another example of a nonclassical manual, one that arises in the study of electrons. When behaviour of electrons was studied in the context of quantum mechanics and relativity theory, it was discovered that it was not sufficient to describe moving electrons in terms of position and momentum. Another variable was required to explain outcomes of the experiments described below.

Suppose we think of an electron as a small round particle that we can move through space. Modern physics has shown that this is not at all a good way to think about an electron, but it is sufficient for us at the moment. If we shoot the electron from a gun and send it on a straight line path, we can deflect it by passing it through a magnetic field. See Figure 3A.5. We can measure the deflection by placing a screen in the path of the electron and recording where the electron hits the screen.

Let us place a two-dimensional coordinate system on the screen with the origin at the spot the electron would have hit if it had not been deflected. We can orient the magnets in such a way that the electron will be deflected only in the y direction, so that it will land at a point $(0, y_1)$ for some positive or negative number y_1. We also can orient the magnets so that the electron will be deflected only in the x direction, landing at a point $(0, x_1)$.

The theory of quantum mechanics requires that with every electron moving through space there is a variable called the spin variable that is associated with how the electron is deflected by magnets. An experiment that measures

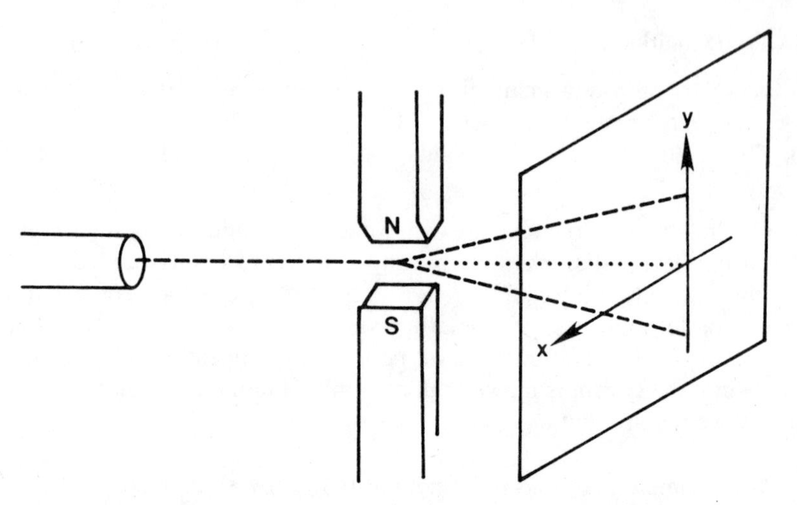

Figure 3A.5. An electron deflection experiment.

this variable is called a Stern–Gerlach experiment, after O. Stern and W. Gerlach, who performed deflection experiments with silver atoms in 1921. While classical properties of spinning bodies were well understood for a long time before there were Stern–Gerlach experiments, the properties of the variable associated with electron spin were anything but classical.

The first nonclassical property of electron spin is that it is "quantized." This means the following. Suppose we send an electron through the magnets and measure its deflection in the y direction, and we observe that the electron arrives at point $(0, y_1)$, with $y_1 > 0$. We say that the *y component of the spin* is positive and of magnitude y_1. Suppose now that we could repeat the experiment exactly, with the same electron moving and spinning in the same way, except that we shall measure the deflection in a direction u different from y. We do this by rotating the magnets a bit. If electron spin were classical, then the component of spin in the u direction would have a magnitude slightly more or less than y_1. Experiments show, however, that the u component of electron spin will have magnitude exactly equal to y_1, and this will be true no matter what the direction u.

We can rephrase the above in terms of manuals. Let us consider the set U of axes on the two-dimensional screen that the electron hits. Then each $u \in U$ determines a direction for a spin measurement, and we denote by E_u the experiment that measures deflection in the direction of the u axis. We say that spin is quantized by saying that every experiment E_u has exactly two outcomes, which we call *u-up* and *u-down*. Then we define the *spin manual* as $\mathscr{S} = \{E_u | u \in U\}$.

3A.13. Project. The manual $\mathscr{S}$ consists of infinitely many pairwise disjoint *dichotomies*, experiments with two outcomes. Verify that $\mathscr{S}$ is a manual and that it is not classical.

Next we consider a second nonclassical property of electron spin. Suppose we prepare experiment E_u to measure the coordinate of spin in the u direction and try to predict which outcomes will occur, u-up or u-down. It seems reasonable to assume that at an instant t_1 before the electron hits the screen, we have no idea what the outcome of experiment E_u will be. In other words, at instant t_1 we can associate with the electron a weight ω_1 on manual $\mathscr{S}$ such that $\omega_1(u\text{-up}) = \omega_1(u\text{-down}) = \frac{1}{2}$. At the instant t_0 the electron hits the screen, a weight ω_0 on manual $\mathscr{S}$ associated with the electron would have to have $\omega_0(u\text{-up}) = 0$ and $\omega_0(u\text{-down}) = 1$ or vice versa ($\omega_0(u\text{-up}) = 1$ and $\omega_0(u\text{-down}) = 0$). In fact, at the same instant t_0, for a direction v different from u, if we could perform E_v simultaneously with E_u, the weight ω_0 would have to have $\omega_0(v\text{-up}) = 1$ and $\omega_0(v\text{-down}) = 0$ or vice versa. To continue our discussion we make the following definition.

3A.14. Definition. A weight on a manual is called *classical* if it has value zero or one on every outcome.

Therefore, a weight on spin manual $\mathscr{S}$ that predicts with certainty (values 0 and 1) the u component of spin for *every* direction u simultaneously at the time the electron hits the screen must be a classical weight on $\mathscr{S}$.

Here is a crucial question: Can we physically prepare an electron so that at some instant t_0 its associated weight ω_0 on the spin manual is classical? From everything physicists have learned about electrons so far, the answer appears to be "no." This is a profound issue at the heart of quantum physics. It is a *principle* that all theories about the physics of an electron will be inconsistent with experimental results if they allow that the electron at some instant can be associated with a classical weight on the spin manual.

What makes this statement of principle profound (and controversial) is the fact that it applies to all theories, even those that have not yet been invented. This is one reason that quantum physics is often a topic of hot debate. It is natural to challenge the principle by asking its proponents how they know that it is not the case that either

(i) the way we now associate electrons with weights on the spin manual is crude, and some day we will discover that the nonclassical weights we observe are really some kind of mixtures of classical weights; or
(ii) this spin manual is crude and some day we will have a better manual, perhaps a classical manual, and a way of associating electron spin with only classical weights on the new manual.

While no one will argue that there is proof that neither (i) nor (ii) will ever happen, all evidence that we have so far appears to point in that direction. We shall return to this debate later.

It is important to emphasize that we constructed our manual to reflect our assumption, based on experimental evidence, that electron spin is nonclassical. We have not deduced the nonclassical nature of spin from the manual. The purpose of the manual in our discussion above is to *describe* a theory of electron spin, not to *predict* it.

Let us now use Hilbert spaces to provide another description of the spin manual $\mathscr{S}$ that will enable us to describe how electrons can be "associated" with weights on $\mathscr{S}$. It will be clear that the electron cannot be associated with a classical weight in this vector space model.

Consider again the set U of axes mentioned above. For each axis u let θ_u be the angle measured counterclockwise between the positive x axis and the u axis. See Figure 3A.6. Now consider a two-dimensional Hilbert space H, and for every $u \in U$ we associate a pair of subspace in H, $E_{K_u} = \{K_u, K_u^\perp\}$, where K_u is the one-dimensional subspace of H oriented at angle $\theta_u/2$ from the positive horizontal axis and $K_u^\perp$ is the orthogonal complement of K_u in H. Refer again to Figure 3A.6. (The reason for this technicality involving angles is explained in Chapter 8.) Be careful not to confuse the axis u, which appears on the physical two-dimensional screen, with the subspace K_u, which lies in the abstract space H.

Clearly, there is a one-to-one correspondence between the spin manual $\mathscr{S}$ and the set $\mathbb{S} = \{E_{K_u} | u \in U\}$. Each experiment $E_u = \{u\text{-up}, u\text{-down}\}$ can be represented by a pair of orthogonal subspaces of H, one subspace representing u-up, the other representing u-down. In Figure 3A.7 we have illustrated experiments E_x and E_y corresponding to the measurement of the coordinate of spin in the x and y directions shown in Figure 3A.5. Notice that in Figure 3A.5 x and y are orthogonal directions, while in Figure 3A.7 the subspaces K_x and K_y make a 45-degree angle. We now explore the consequences of that angle.

Physicists use matrix equations to describe electron spin. Those equations suggest properties attributed to electrons. We shall not go into the details of

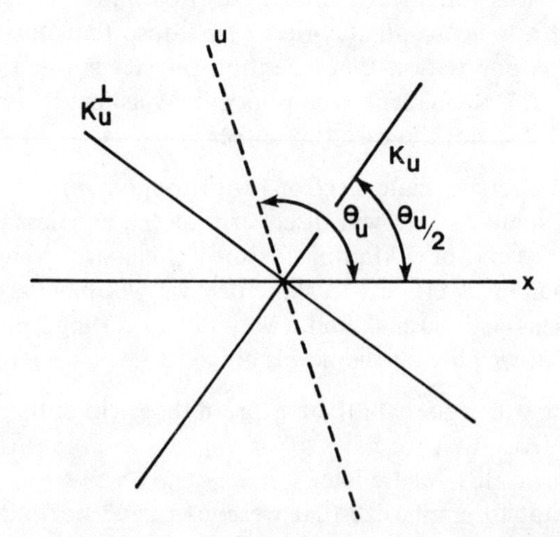

Figure 3A.6. An orthogonal pair of subspaces, Ku and $Ku^\perp$ representing an experiment to measure electron spin deflection along the u axis.

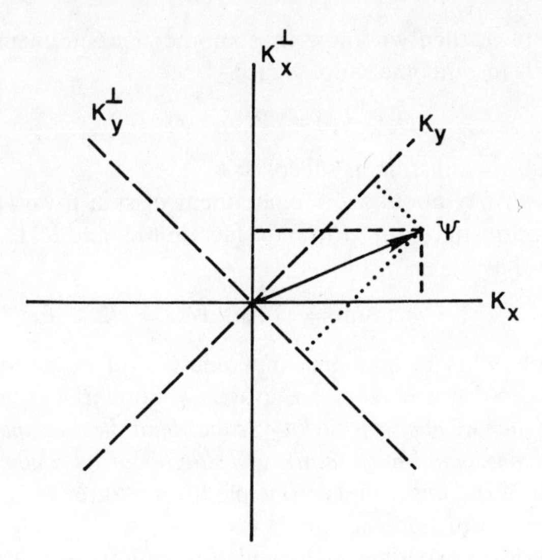

Figure 3A.7. The probabilities of the occurrence of outcomes x-up and y-up are calculated using the projections of state vector ψ onto the subspaces K_x and K_y.

the equations, but we shall use the spin manual to state the property that is of interest to us: An electron in a Stern–Gerlach apparatus is said to exist at every instant in time in a "state" ψ, which can be represented by a unit vector in the Hilbert space H of Figure 3A.7. With each state ψ is associated a weight on the spin manual $\mathscr{S}$ as follows: for $u \in U$, $\omega(u\text{-up})$ is the square of the length of the projection of ψ onto the subspace K_u, and $\omega(u\text{-down})$ is the square of the length of the projection of ψ onto the subspace $K_u^{\perp}$. (We shall consider projections more generally in Chapter 4, but we are assuming here that you have already encountered projections when you studied finite dimensional linear algebra.)

We write

$$\begin{aligned} \omega_\psi(u\text{-up}) &= \| \mathrm{Proj}_{K_u} \psi \|^2, \\ \omega_\psi(u\text{-down}) &= \| \mathrm{Proj}_{(K_u^{\perp})} \psi \|^2. \end{aligned} \tag{3A.1}$$

Thus, if the electron is in state ψ, then we know that the probability of obtaining the outcome "x-up" if we do experiment E_x is the square of the projection of state vector ψ onto subspace K_x.

3A.15. Project. Show that ω given by equation (3A.1) is a weight on $\mathscr{S}$ and that it is a nonclassical weight.

This description of the spin states of an electron raises some delicate questions about measurements. Suppose first that we measure for the x component of spin and obtain outcome x-up. The electron is then in state ψ. If we assume that we can let the electron pass through the screen and remain in state ψ, then we know that another measurement of x-spin will certainly result in outcome x-up. Hence,

$$\omega_\psi(x\text{-up}) = \| \mathrm{Proj}_{K_x} \psi \|^2 = 1. \tag{3A.2}$$

This means that ψ must lie in subspace K_x.

What can we say about the y component of spin if we *know for certain* that the x component is x-up? From the fact that K_x and K_y make a 45-degree angle, we see that

$$\omega_\psi(y\text{-up}) = \| \mathrm{Proj}_{K_y} \psi \|^2 = 1/2. \tag{3A.3}$$

Thus, the probability of obtaining outcome y-up if we perform experiment E_y when the electron is in state ψ is exactly $\frac{1}{2}$. Similarly, $\omega_\psi(y\text{-down}) = \frac{1}{2}$. In other words, *when we have certain knowledge about the x component of electron spin, we have maximally uncertain knowledge about its y component.* This is an expression of the uncertainty principle for measurements of components of electron spin in orthogonal directions.

Can we perform experiments E_x and E_y simultaneously? Not if the spin manual and the allowable weights on it are determined by our Hilbert space model above. In addition to the fact that the spin manual $\{E_x, E_y\}$ consisting of just these two experiments is nonclassical, we see that *no state vector ψ can induce a classical weight on it.*

Even if we cannot perform experiments E_x and E_y simultaneously, can we at least say that the electron "has" components of spin in both x and y directions simultaneously? Even that appears impossible. In a famous paper published in 1935 by Einstein, Podolsky, and Rosen (called the EPR paper), the authors argued that the only way quantum mechanics can be logically consistent is if it prohibits the possibility that an electron "has" a coordinate of spin in a given direction *until* it is measured. They based their argument on an imaginary experiment (called an EPR experiment) that they dreamed up to follow quantum mechanics to its logical conclusion. They concluded that because quantum mechanics cannot account for an element of "reality," electron spin before it is measured, it must be an "incomplete" theory. We will consider the EPR experiment in detail in Chapter 8.

The debate over the meaning of "reality" and "completeness" continues today. Extensive current research in mathematics as well as theoretical and experimental physics concerns this debate. Delicate EPR experiments carried out in France in 1985 appear to justify the logical consistency argument of Einstein, Podolsky, and Rosen, but this does not settle the debate. Einstein et al. might have been right in their argument that quantum mechanics must hold that certain properties do not exist until they are measured, but who is to say that this is not a fact of nature? It seems that quantum mechanics, as long as it contains this tenet, will never be free from attack by the philosophically inclined.

Before we move on to logics, let us consider a Hilbert space model for the bow tie manual. Let H be three-dimensional real vector space. Let

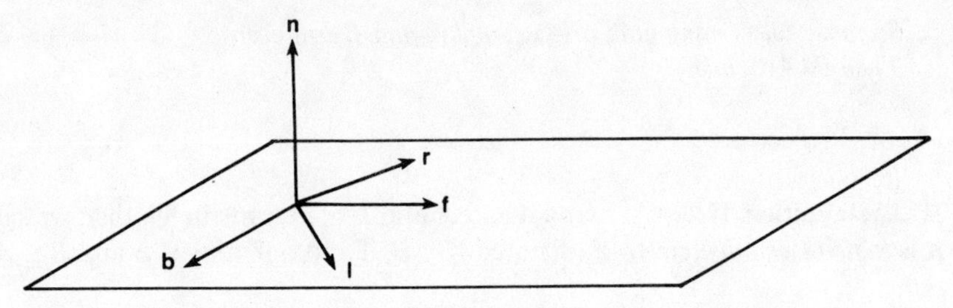

Figure 3A.8. A Hilbert space model for the bowtie manual.

$E=\{l,r,n\}$ and $F=\{b,f,n\}$ be orthonormal bases for H as shown in Figure 3A.8.

It is easy to see that the manual $\mathfrak{M}=\{E,F\}$ has the orthogonality diagram of Figure 3A.2. (Be careful about the different meanings of "orthogonal," as we mentioned in Example 3A.4C.) For each unit vector $\psi\in H$ define a weight on $\mathfrak{M}$ by

$$\omega_\psi(w)=|\langle\psi,w\rangle|^2 \quad \text{for all } w\in E\cup F. \tag{3A.4}$$

Then this set of weights is a proper subset of $\Omega_{\mathfrak{M}}$, the set of all weights on $\mathfrak{M}$.

3A.16. Project. Show that $\Omega_{\mathfrak{M}}$ contains more than one classical weight but that there is exactly one classical weight $\omega\in\Omega_{\mathfrak{M}}$ for which there exists unit vector $\psi\in H$ satisfying equation (3A.4) for all $w\in E\cup F$.

In Part B of this chapter we consider further connections between weights, states, and unit vectors.

Part B: Logics and State Functions

We can think of an event A, a subset of experiment E in a manual $\mathfrak{M}$, as the question: "If we perform experiment E, will A occur?" Consider, for example, events $A=\{l\}$ and $B=\{f,b\}$ in the bow tie manual of Figure 3A.1. If we test for A by performing E and A occurs, then we know that $\{n\}$ did not occur, so if we had performed F instead of E, we know that B *would have occurred.* In this sense we say "A implies B." More generally, we make the following definition.

3B.1. Definition. If $\mathfrak{M}$ is a manual and A and B are events in $\mathfrak{M}$, then we say *A implies B*, denoted $A\leqq B$, if and only if there is an event C with $C\perp A$ and $C\cup A \operatorname{op} B$.

3B.2. Lemmas.

A. *Let $\mathfrak{M}$ be a manual. Then implication is a partial ordering on the collection of events in $\mathfrak{M}$.*

B. *Suppose $\mathfrak{M}$ is a manual, $\omega \in \Omega_{\mathfrak{M}}$, and A and B are events in $\mathfrak{M}$ with $A \leqq B$. Then $\omega(A) \leqq \omega(B)$.*

PROOF. Project. □

3B.3. Definition. If $\mathfrak{M}$ is a manual and A and B are events in $\mathfrak{M}$, then we say A is *logically equivalent* to B, denoted $A \leftrightarrow B$, if and only if $A \leqq B$ and $B \leqq A$.

Clearly, $\leftrightarrow$ is an equivalence relation on the set of events in a manual.

3B.4. Lemmas. *Suppose $\mathfrak{M}$ is a manual.*

A. *If A, B are events in $\mathfrak{M}$, then $A \operatorname{op} B$ if and only if $A \leftrightarrow B$.*
B. *If $E, F \in \mathfrak{M}$, then $E \leftrightarrow F$.*

PROOF. Project. □

3B.5. Lemmas. *Suppose $\mathfrak{M}$ is a manual.*

A. *If A, B, C, D are events in $\mathfrak{M}$ and $A \leftrightarrow C$ and $B \leftrightarrow D$, then $A \leqq B$ if and only if $C \leqq D$.*
B. *If A is an event in $\mathfrak{M}$ and $E, F \in \mathfrak{M}$, with $A \subseteq E$ and $A \subseteq F$, then $F \backslash A \leftrightarrow E \backslash A$.*

PROOF. Project. □

3B.6. Definitions. Suppose $\mathfrak{M}$ is a manual.

A. If A is an event in $\mathfrak{M}$, then we define

$$\underline{[A]} = \{B \mid B \text{ is an event in } \mathfrak{M} \text{ and } A \leftrightarrow B\},$$

and we call $[A]$ the *logical proposition determined by A*. We say we *test for proposition* $[A]$ if we test for any event in $\mathfrak{M}$ logically equivalent to A. An event used to test for $[A]$ confirms that $[A]$ *is true* (respectively, *false*) if the event occurs (respectively, does not occur). For outcome x we write $[x]$ for $[\{x\}]$.
B. We define $\underline{\Pi(\mathfrak{M})} = \{[A] \mid A \text{ is an event in } \mathfrak{M}\}$.
C. We define a partial order on $\Pi(\mathfrak{M})$ by

$$\underline{[A] \leqq [B]} \text{ if and only if } A \leqq B \text{ in } \mathfrak{M},$$

in which case we say $[A]$ *implies* $[B]$.
D. For $[A] \in \Pi(\mathfrak{M})$, we define the *orthocomplement* of $[A]$ by $[A]' = [E \backslash A]$, where E is any experiment in $\mathfrak{M}$ with $A \subseteq E$.
E. If $[A], [B] \in \Pi(\mathfrak{M})$, we say that they are *orthogonal*, denoted by $[A] \perp [B]$, if and only if $[A] \leqq [B]'$.
F. The set $\Pi(\mathfrak{M})$, together with the implication $\leqq$ and orthocomplementation is called the *operational logic* of manual $\mathfrak{M}$.

Lemmas 3B.4 show that the partial order and the orthocomplementation on $\Pi(\mathfrak{M})$ are well defined.

3B.7. Examples and Projects.

A. Let $\mathfrak{M}$ be a manual consisting of just one dichotomy $E = \{x, y\}$. Find the logic $\Pi(\mathfrak{M})$.
B. Show that if $[A] \leqq [B]$, then a test for $[A]$ which confirms that $[A]$ is true at the same time confirms that $[B]$ is true. That is why we use the word "implies" to express the relation $[A] \leqq [B]$.
C. Show that $[A] \leqq [B]'$ if and only if $[B] \leqq [A]'$.

3B.8. Lemma. *For events A, B in manual $\mathfrak{M}$, $[A] \perp [B]$ if and only if $A \perp B$.*

PROOF. Project. □

As we shall see in Chapter 4, if $\mathscr{F}(H)$ is the frame manual for a Hilbert space H, then its operational logic can be identified with the structure of subspaces of H, which historically was the prototype for a "quantum logic."

We have motivated the notion of an operational logic by considering physical experiments, and we have seen a sharp distinction between classical and nonclassical manuals. Next we shall study logics in greater generality to see how differences between classical and nonclassical physics manifest themselves in logics.

3B.9. Definition. Let P be a set with a partial ordering $\leqq$. Then $(P, \leqq)$ is called a *lattice* if for all $p, q \in P$ the set $\{p, q\}$ has a greatest lower bound and a least upper bound in P.

We call glb$\{p, q\}$ the *meet* of p and q, and we denote it by $p \wedge q$.
We call lub$\{p, q\}$ the *join* of p and q, and we denote it by $p \vee q$.

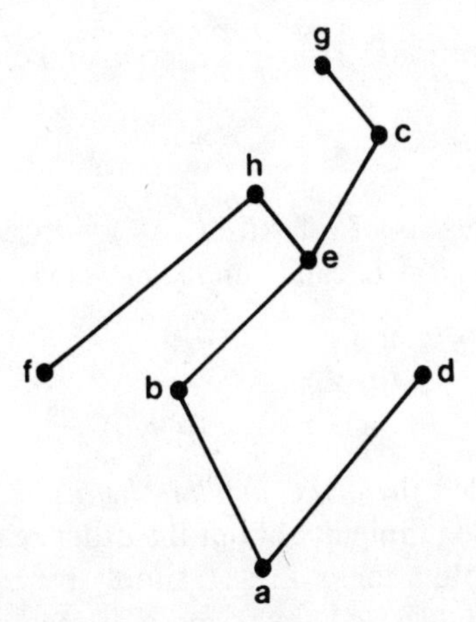

Figure 3B.1. A tree diagram for a lattice.

For a finite lattice it is often helpful to draw a "tree" diagram, as in Figure 3B.1, for example. We draw p below q and connect them with a line segment if $p \leqq q$ and if there does not exist w with $p < w < q$. By the fact that the partial order is transitive, we read right from the diagram that $m \leqq n$ if we can find a path that moves steadily upward from m to n. So in the figure we see that $a \leqq h$ and $a \leqq d$, but $d \nleqq e$ and $e \nleqq d$. We also see immediately that $g \wedge h = e$ and $b \vee f = h$.

3B.10. Definitions. Let $(L, \leqq)$ be a lattice.

A. A *unit* 1_L for L is a member of L such that $p \leqq 1_L$ for all $p \in L$.
B. A *zero* 0_L for L is a member of L such that $0_L \leqq p$ for all $p \in L$.

As usual, we drop the subscript L for the unit and zero if there is no ambiguity about which lattice is meant.

3B.11. Examples and Project.

A. Let X be a nonempty set and let $\text{Sub}(X)$ be the collection of all subsets of X. Show that $\text{Sub}(X)$, partially ordered by set inclusion [$p \leqq q$ if and only if $p \subseteq q$], is a lattice with unit and zero. Draw the tree diagram for such a lattice in the case X has exactly three members.
B. Let H be a finite dimensional Hilbert space, and let L be the collection of all subspaces of H. Define a partial order on L as follows: for $R, S \in L$, $R \leqq S$ if and only if R is a subspace of S. Show that $(L, \leqq)$ is a lattice with unit and zero. Draw the tree diagram for L in the case H has dimension 2. Later we shall extend this idea to infinite dimensional Hilbert spaces.
C. Let $\mathfrak{M}$ be the bow tie manual. Draw the tree diagram for $(\Pi(\mathfrak{M}), \leqq)$, and determine whether or not it is a lattice. Does it have unit and/or zero?

Motivated by operational logics, we now make the following general definitions.

3B.12. Definitions.
A. A *logic* $(L, \leqq, ')$ consists of a lattice $(L, \leqq)$ with unit and zero, together with an operation $' : L \to L$, called an *orthocomplementation*, that satisfies:

(i) for all $p \in L$, $p'' = p$ and $p \wedge p' = 0_L$;
(ii) for $p, q \in L$, if $p \leqq q$, then $q' \leqq p'$;
(iii) for $p, q \in L$, if $p \leqq q$, then $q = p \vee (p' \wedge q)$.

Property (iii) is called the *orthomodular identity*.

Where there is no ambiguity about the order relation or orthocomplementation, we denote a logic $(L, \leqq, ')$ simply by L.
B. If L is a logic and $p, q \in L$, we say p is *orthogonal* to q, denoted $p \perp q$, if and only if $p \leqq q'$.

The members of L are called *propositions*. For example, if $\mathfrak{M}$ is a manual and A is an event in $\mathfrak{M}$, then the proposition $p = [A]$ is the proposition: "If we test for event A, it will occur." Naturally, this proposition may be true at some instants of time and false at others. In our applications we shall interpret the proposition p' as the negative of proposition p. Thus, p is true precisely when p' is false and vice versa.

The relation $p \leqq q$ will be interpreted as the statement: "Proposition p implies proposition q." The propositions $p \wedge q$ and $p \vee q$ will be interpreted respectively as the conjunction "p and q" and the disjunction "p or q." When we consider states on a logic we shall hypothesize that the proposition 1_L is true for all states and that the proposition 0_L is true for no state.

There are systems $(L, \leq, ')$ that satisfy all the properties of logics except that they are not lattices. If we were to explore logics in more detail, we would use a more general definition of a logic by requiring only that $(L, \leqq)$ be a partially ordered set and by replacing property (iii) with a slightly weaker version. In nearly all of our examples, however, our operational logics will be lattices. Moreover, the subspaces of a Hilbert space form a lattice that is the logic of orthodox quantum mechanics, and we shall study this in great detail. That is why we use a slightly less general definition of a logic than you might find elsewhere.

Since lattices of subspaces of Hilbert space satisfy more properties than more general logics, it is important to try to determine which of the properties of a logic are really necessary to describe physical nature and which properties are superfluous, perhaps even misleading, to theoretical physicists. There is currently much research being done by mathematicians and physicists trying to make these determinations. Perhaps you will have some ideas after we study logics a little further.

3B.13. Theorem (DeMorgan's Law). *If L is a logic and $p, q \in L$, then*

$$(p \vee q)' = p' \wedge q' \quad \textit{and} \quad (p \wedge q)' = p' \vee q'.$$

PROOF. Project. □

Next we consider compatibility in logics.

3B.14. Definition.

A. Two propositions p and q in logic L are called *compatible* if there exist $u, v, w \in L$ such that:

(i) $\{u, v, w\}$ is a pairwise orthogonal set in L;
(ii) $u \vee v = p$ and $v \vee w = q$.

We call $\{u, v, w\}$ a *compatibility decomposition* for p and q.

B. A *quantum logic* is a logic with at least two propositions that are not compatible.

C. A *classical logic* is a logic in which every pair of propositions is a compatible pair.

3B.15. Theorem. *If $\{u, v, w\}$ is a compatibility decomposition for p and q in a logic L, then*

(i) $u = p \wedge q'$;
(ii) $w = p' \wedge q$;
(iii) $v = p \wedge q = (p' \vee q) \wedge p = (p \vee q') \wedge q$.

Note that this implies that u, v, and w are uniquely determined by p and q.

PROOF. Project.* □

3B.16. Theorem. *If p and q are propositions in a logic L, then p and q are compatible if and only if*

(i) $p = (p \wedge q) \vee (p \wedge q')$, *and*
(ii) $q = (q \wedge p) \vee (q \wedge p')$.

PROOF. Project. □

Notice that it is always true that $(p \wedge q) \vee (p \wedge q') \leqq p$. Theorem 3B.16 then states that we generally do not have that $p \leqq (p \wedge q) \vee (p \wedge q')$ unless p and q are compatible. In other words, knowing that proposition p is true is not sufficient for concluding that at least one of the following is true:

(i) *p and q are simultaneously true, or*
(ii) *p and "not-q" are simultaneously true.*

This logical structure captures the idea that for some physical systems there might exist a pair of propositions whose truth values simply cannot be simultaneously determined. Theorem 3B.16 is the manifestation in the logic of the notion that pairs of events in a manual might not be simultaneously testable.

3B.17. Theorem. *If A and B are compatible events in manual $\mathfrak{M}$, then $[A]$ and $[B]$ are compatible in the operational logic $\Pi(\mathfrak{M})$.*

PROOF. Project. □

In later chapters we show that we can associate the subspaces of a Hilbert space H with propositions in a logic, and certain linear operators on H with physical observables. Then we show that a pair of operators (observables) *commutes* if and only if a certain collection of subspaces (propositions) is *pairwise compatible*. This neatly ties up the notions of simultaneously testable observables, commuting operators, and logically compatible propositions in the Hilbert space model for quantum physics.

3B.18. Definition. Suppose L is a logic. We define a *state* on L as a function $s: L \to [0, 1]$ such that

(i) for $p, q \in L$, if $p \perp q$, then $s(p \vee q) = s(p) + s(q)$, and
(ii) $s(1_L) = 1$.

If $\mathfrak{M}$ is a manual with operational logic $L = \Pi(\mathfrak{M})$, then a state s on L is also called a *probability measure* on L. For if $[A]$ is a member of L, then $s([A])$ is usually interpreted as *the probability that the event A (or an equivalent event) will occur if tested while the system represented by* $\mathfrak{M}$ *is in state s.* That is why we require that if A and B are orthogonal events, the probability of the occurrence of their join is the sum of their individual probabilities of occurring. In this sense their join is interpreted as the disjunction: "either A or B occurs."

3B.19. Example and Project. Let $\mathfrak{M}$ be the bow tie manual, and suppose ω is a weight on $\mathfrak{M}$. Let $s: \Pi(\mathfrak{M}) \to [0, 1]$ be defined by $s([A]) = \omega(A)$. Show that s is a state on $\Pi(\mathfrak{M})$. We call s the *state on* $\Pi(\mathfrak{M})$ *induced by* ω.

The notion of a state on a logic provides us with another way to examine the fundamental uncertainty of nature.

3B.20. Lemma. *Suppose L is a logic and* $\{s_1, \ldots, s_n\}$ *is a finite collection of states on L. Suppose* $\{\alpha_1, \ldots, \alpha_n\}$ *is a collection of real numbers satisfying:*

(*i*) *for all* $k = 1, \ldots, n$, $0 \leqq \alpha_k \leqq 1$;
(*ii*) $\sum_{k=1}^{n} \alpha_k = 1$.

Then the function $s = \sum_{k=1}^{n} \alpha_k s_k$, *defined for every* $p \in L$ *by*

$$s(p) = \sum_{k=1}^{n} \alpha_k s_k(p) \quad \text{is a state on } L.$$

We call s a mixture of the states $s_1, \ldots, s_n$, *and we refer to a set of numbers satisfying* (*i*) *and* (*ii*) *as a mixing set of numbers.*

PROOF. Project. □

3B.21. Definition. A state s on logic L is called *dispersion-free* if, for all $p \in L$, either $s(p) = 0$ or $s(p) = 1$.

Dispersion-free states are also called *classical states.*

A key question in the foundation of physics is: When is a state that is not dispersion-free a mixture of dispersion-free states? For example, in a physical experiment we might be able to determine that at a certain time a physical system is in a state that is a mixture $s = \sum_{k=1}^{n} \alpha_k s_k$ of dispersion-free states $s_1, \ldots, s_n$. Then we say s is a mixture reflecting an *epistemic uncertainty.* By that we mean that we believe the system is in one of the dispersion-free states,

s_k, except that we do not know which one. We know only the probability α_k that the system is in state s_k.

Next we turn our attention to a more profound uncertainty—a state that is not dispersion-free and *cannot be written as a mixture of dispersion-free states.* We examine this situation with an example, which we consider after we make the following definition.

3B.22. Definition. A *pure state* on a logic L is one that cannot be written as a nontrivial mixture of other states on L.

We consider a firefly again. This time we put our firefly in a five-chamber box, pictured in Figure 3B.2. The firefly is free to roam among the five chambers and to light up at will. The sides of the box are windows with

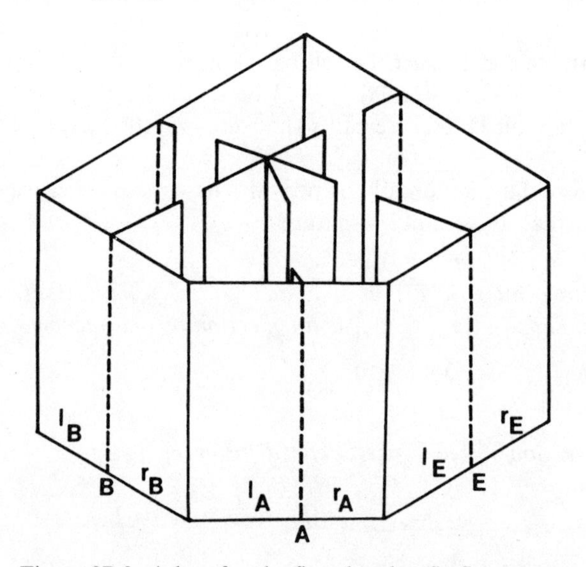

Figure 3B.2. A box for the five-chamber firefly system.

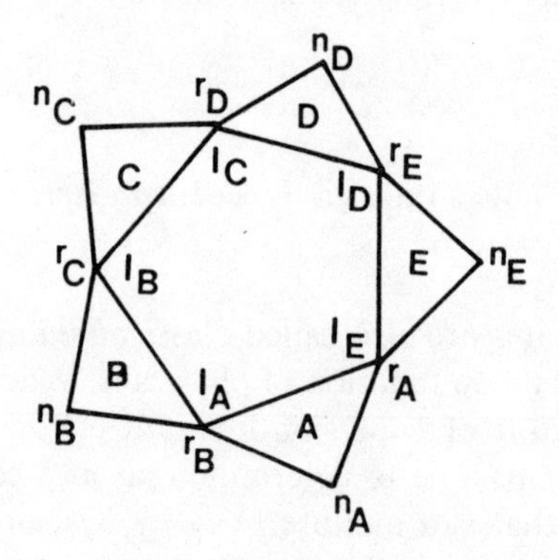

Figure 3B.3. The pentagon manual.

vertical lines down their centers. We have five experiments—five windows to look at. For each experiment E we record l_E, r_E, or n_E if we see, respectively, a light to the left of the center line, a light to the right, or no light. In Figure 3B.3 we have drawn an orthogonality diagram for this manual of experiments.

3B.23. Project. Explain why we have identified $l_E = r_A$ but have not identified $n_A = n_E$.

Let us make a chart of all the weights describing the condition of the firefly based on our knowledge of "reality."

Firefly lit in chamber:	$l_A = r_B$	$l_B = r_C$	$l_C = r_D$	$l_D = r_E$	$l_E = r_A$	n_A	n_B	n_C	n_D	n_E
1: $\omega_1 \to$	1	0	0	0	0	0	0	1	1	1
2: $\omega_2 \to$	0	1	0	0	0	1	0	0	1	1
3: $\omega_3 \to$	0	0	1	0	0	1	1	0	0	1
4: $\omega_4 \to$	0	0	0	1	0	1	1	1	0	0
5: $\omega_5 \to$	0	0	0	0	1	0	1	1	1	0
Firefly not lit	0	0	0	0	0	1	1	1	1	1
$\omega \to$	$\frac{1}{2}$	$\frac{1}{2}$	$\frac{1}{2}$	$\frac{1}{2}$	$\frac{1}{2}$	0	0	0	0	0

Now consider the weight ω defined on $\mathfrak{M}$ as follows:

$$\left.\begin{array}{l}\text{(i) } \omega(n_A) = \omega(n_B) = \omega(n_C) = \omega(n_D) = \omega(n_E) = 0, \\ \text{(ii) } \omega(x) = \tfrac{1}{2} \quad \text{for all other outcomes } x.\end{array}\right\} \qquad (*)$$

We have listed ω in the last row of the chart above.

The operational logic $L = \Pi(\mathfrak{M})$ is a lattice. Although it is not trivial to prove, it is true that the state s on L induced by ω [i.e., $s([A]) = \omega(A)$] is not a mixture of any collection of dispersion-free states on L. In fact, it is a pure state and cannot be written as a mixture of any collection of states on L. From the chart above, we can see at least that ω is not a mixture of any of the dispersion-free weights [weights having value 0 or 1 only] in the chart. In Example 6A.16 we shall investigate more thoroughly a pure state that is not dispersion-free.

We refer to a pure state that is not dispersion-free as a state of *ontological uncertainty* as opposed to epistemic uncertainty. While the latter reflects an uncertainty in *our knowledge* about which dispersion-free state the system is in, the former reflects a fundamental uncertainty in *nature* that is independent of physical theories invented by humans. It is not surprising that some people are reluctant to accept a theoretical foundation of quantum mechanics based on ontological uncertainties. Faith in the inevitable triumph of science usually allows uncertainty only as a temporary, incomplete description of a physical system, not as a fundamental state of nature.

Consider the ontological uncertainty in the five-chamber firefly system. What is the "real" physical state corresponding to the weight ω that assigns $\frac{1}{2}$ to each of the outcomes corresponding to a light-up? It is hard for us to imagine how a firefly could behave to produce such a state, tempting us to say that no such state can exist in "reality"—it is only a mathematical quirk. But we must be careful before dismissing such a state as "unreal." Perhaps in this state the firefly can sense our presence and fly immediately to whatever chamber we look at and light up, half the time on one side of the center line, half the time on the other. This is a stage-struck firefly! Surely we have made a strained explanation. What it illustrates is a desperate attempt to fit the mechanical system (the firefly in the box) to the theoretical model (the logic and its states).

In physics it is often the case that the theoretical model comes first from manipulations of equations, and the mechanical system (an "antiparticle," for example) is then described to fit the theoretical model. When strained explanations are required to fit the two, physicists are faced with a choice: Should they change the mechanical model or the theoretical model? The final judgements are determined by experiments. The larger the number of experiments that are consistent with the theoretical model, the more faith people have in the theory, and then pressure builds to change the mechanical model. Of course, if an experiment produces results inconsistent with a theory, that theory is immediately changed or discarded, no matter how nicely it fits with a mechanical model.

What role does ontological uncertainty play in the evaluation of a theoretical model? If a model containing an ontological uncertainty fits well with all experiments known, it seems unwise to throw it out on the basis of a philosophical prejudice against ontological uncertainty, although one might cling to the hope that the model eventually will be adjusted to contain only epistemic uncertainties. How tenaciously to cling to such hope is a personal decision.

Fortunately, it is not necessary for a physicist or a mathematician to take sides in the debate about the existence of ontological uncertainty in nature. We can simply provide the framework in which the issues can be precisely and unambiguously stated and leave the debate to others. That is what we shall do in the rest of this book.

CHAPTER 4

Subspaces in Hilbert Space

In Chapter 3 we used two-dimensional Hilbert space to model a manual for measuring electron spin, because every experiment had only two outcomes: spin-up and spin-down. Of course, many physical experiments have outcome sets that are not finite. A measurement of energy, position, or momentum of a moving particle usually means obtaining an outcome from an infinite number of possibilities. In this chapter we shall learn about infinite dimensional Hilbert spaces. In particular, we shall consider the subspace structure of infinite dimensional Hilbert spaces. These provide the logics for orthodox quantum physics.

Throughout this chapter, unless otherwise noted, H will denote an arbitrary Hilbert space with a countable basis. As usual, $\mathfrak{R}$ and $\mathfrak{C}$ denote, respectively, the real and complex numbers.

4.1. Definition. A set $M \subseteq H$ is a *linear manifold* in H if and only if for all $x, y \in M$ and all $\alpha \in \mathfrak{C}$, $x + y$ and αx both belong to M.

You probably recognize that if H is finite dimensional, a linear manifold is just a "vector subspace." In the infinite dimensional case, however, the word "subspace" is reserved for certain kinds of manifolds.

4.2. Definition. A set $S \subseteq H$ is *closed* if every Cauchy sequence in S converges in norm to a vector in S.

4.3. Theorem. *If $\mathbb{S}$ is a collection of closed subsets of H, then $\cap \mathbb{S}$ is a closed set in H.*

PROOF. Project. □

4.4. Definition. If $S \subseteq H$, we define the *closure of* S by

$$\text{clos}(S) = \cap\{C \mid S \subseteq C, \text{ and } C \text{ is closed}\}.$$

Observe from Theorem 4.3 that clos(S) is a closed set. From the definition of intersection, observe that it is a subset of every closed set containing S. An alternative definition of the closure of S is

$$\text{clos}(S) = \{x \mid \text{there exists sequence } \langle\langle x_k \rangle\rangle \text{ in } S \text{ that converges in norm to } x\}.$$

It is not difficult to prove that the two definitions are equivalent, but we shall omit the proof.

4.5. Example and Project. Let H be a two-dimensional Hilbert space, and let $S = \{x \in H \mid \|x\| < 1\}$. Suppose $0 \neq x \in S$, and consider the sequence defined by $x_k = (1/\|x\| - 1/k)x$ for all $k \in \mathbb{N}$. Show that $x_k \in S$ for all $k \in \mathbb{N}$ and that $\langle\langle x_k \rangle\rangle$ converges in norm to $x/\|x\|$, which is not a member of S. Thus, S is not closed.

4.6. Definition. A *subspace* in a Hilbert space H is a closed linear manifold in H.

Observe that if K is a subspace in H, then K is a Hilbert space.

4.7. Examples and Project. (Example A contains a project, Example B does not.)

A. Let $H = l^2$. (See Example 2.3B.) Let $M = \{x \in H \mid x = \langle\langle x_k \rangle\rangle$, and all but finitely many x_k equal zero$\}$. Show that M is a linear manifold in H that is not closed. (This project is at level $*$.)

B. Suppose $[a, b]$ is an interval in $\mathfrak{R}$. Then the following are linear manifolds in $\mathscr{L}_2(a, b)$. (See Example 2.3C.)

$$C[a, b] = \{\psi \mid \psi: [a, b] \to \mathfrak{C} \text{ and } \psi \text{ is continuous on } [a, b]\}.$$

$$C^\infty[a, b] = \{\psi \mid \psi: [a, b] \to \mathfrak{R} \text{ and } \psi \text{ has derivatives of all orders on } (a, b) \text{ and is continuous at } a \text{ and } b\}.$$

Be cautioned that our notation obscures the fact that both sets above are sets of equivalence classes, where ψ_1 is equivalent to ψ_2 if and only if $\psi_1 = \psi_2$ μ-ae on $[a, b]$, where μ is Lebesgue measure.

If ψ is continuous on $[a, b]$, then $|\psi|^2$ is continuous, hence integrable on $[a, b]$. From this and the fact that functions are continuous at all points at which they are differentiable, we have that

$$C^\infty[a, b] \subseteq C[a, b] \subseteq \mathscr{L}_2(a, b).$$

That $C^\infty[a, b]$ and $C[a, b]$ are linear manifolds in $\mathscr{L}_2(a, b)$ follows from standard theorems in calculus. Although we shall not prove it here, neither of these manifolds is closed in $\mathscr{L}_2(a, b)$. On the other hand, the closure of $C[a, b]$ is $\mathscr{L}_2(a, b)$.

4.8. Theorem. *If K is a subspace in H and B is a basis for K, then there is a basis B_0 for H with $B \subseteq B_0$.*

PROOF. B is an orthonormal set in H, so is contained in a maximal orthonormal set, which is countable by Theorem 2.26. □

4.9. Lemmas.

A. *If K is a subspace in H, then there exists basis B for K such that for all $x \in H$, $x \in K$ if and only if $x = \sum_{b \in B} \langle x, b \rangle b$. (This project is at level * because it requires the use of Zorn's lemma.)*
B. *If K is a subspace in H and B is a basis for K, then for all $x \in H$, if $x \perp b$ for all $b \in B$, then $x \perp y$ for all $y \in K$.*

PROOF. Project. □

4.10. Theorem. *If M is a linear manifold in H, then* clos(M) *is a linear manifold in H and is, therefore, a subspace in H.*

PROOF. That clos(M) is a manifold follows from Theorem 2.21B. □

4.11. Definition. If $S \subseteq H$, we define the *span of S* by

$$\underline{\vee S} = \cap \{K \mid K \text{ is a subspace in } H \text{ with } S \subseteq K\}.$$

4.12. Theorem. *If $S \subseteq H$, then*

(i) $\vee S$ *is a subspace in H;*
(ii) $S \subseteq \vee S$;
(iii) for all subspaces K in H, if $S \subseteq K$, then $\vee S \subseteq K$.

In other words, $\vee S$ is the smallest subspace in H containing S.

PROOF. Project. □

4.13. Theorem. *Suppose $S \subseteq H$ and $M = \{\sum_{k=1}^{n} \lambda_k x_k \mid n \in \mathbb{N}, \lambda_k \in \mathfrak{C}, x_k \in S\}$. Then*

(i) M is a linear manifold in H;
(ii) $\vee S =$ clos(M).

In other words, the span of S is the closure of the set of all finite linear combinations of members of S.

PROOF. Project. □

4.14. Corollary. *If $S \subseteq H$, then $x \in \vee S$ if and only if there is a sequence $\langle\langle x_k \rangle\rangle$ in S and a sequence $\langle\langle \lambda_k \rangle\rangle$ in $\mathfrak{C}$ with $x = \sum_{k=1}^{\infty} \lambda_k x_k$. If the last equality holds, we say x is a linear combination of the members of S.*

PROOF. This is immediate from Theorem 4.13. □

Now we are ready to make the connection between the subspace structure for a Hilbert space and quantum logics.

4.15. Definitions.

A. If $S \subseteq H$, we define $S^{\perp} = \{x \in H \mid \text{for all } s \in S, x \perp s\}$. We call $S^{\perp}$ the *orthogonal complement of S*.

B. If $\mathbb{S}$ is a collection of subsets of H, we write

$$\bigvee_{S \in \mathbb{S}} S \quad \text{for } \vee(\cup \mathbb{S}).$$

4.16. Lemmas.

A. If $S \subseteq H$, then S is a countable orthonormal set if and only if S is a basis for $\vee S$.

B. If $\varnothing \subset A \subset B \subseteq H$ ($\varnothing \neq A \neq B$) and B is a basis for H, then $(B \backslash A)^{\perp} = \vee A$.

PROOF. Project. □

4.17. Theorem. *If $S \subseteq H$, then*

(i) $S^{\perp} \cap S = \{0\}$;

(ii) $S^{\perp}$ *is a subspace in H (even if S is not);*

(iii) if $S \subseteq T \subseteq H$, *then* $T^{\perp} \subseteq S^{\perp}$;

(iv) $S \subseteq (S^{\perp})^{\perp}$;

(v) if S is a subspace in H, then $(S^{\perp})^{\perp} = S$;

(vi) if $\mathbb{S}$ is a collection of subspaces in H, then $(\vee_{S \in \mathbb{S}} S)^{\perp} = \bigcap_{S \in \mathbb{S}} S^{\perp}$ *and* $\vee_{S \in \mathbb{S}} S^{\perp} = (\bigcap_{S \in \mathbb{S}} S)^{\perp}$;

(vii) if S is a countable, orthonormal set, then $S^{\perp} \cap \vee S = \{0\}$.

PROOF. Project.* □

4.18. Theorem. *If H is a Hilbert space and $\mathbb{L}$ is the collection of all subspaces in H, then $\mathbb{L}(H) = (\mathbb{L}, \subseteq, ^{\perp})$ is a logic with $0 = \{0\}$ and $1 = H$*

PROOF. If $K_1, K_2 \in \mathbb{L}$, then $\{K_1, K_2\}$ has least upper bound $K_1 \vee K_2$ and greatest lower bound $K_1 \bigcap K_2$. So $(\mathbb{L}, \subseteq)$ is a lattice. Clearly, $\{0\}$ and H are, respectively, the least and greatest members of the lattice. It remains to show that $^{\perp}$ is an orthocomplementation. Theorem 4.17, however, gives us everything we need except the orthomodular identity, which we leave as a project (at level *). □

In example 3A.4C we saw that if H is a Hilbert space, then the collection $\mathcal{F}(H)$ of orthonormal bases for H is a manual, which we called the frame manual for H. Then in that same chapter, in a remark following Example 3B.8, we promised that we would eventually link the logic of a frame manual with the subspace structure of the Hilbert space. We fulfill that promise now.

Let us begin by reviewing the definitions associated with a frame manual $\mathfrak{M} = \mathscr{F}(H)$. The events in $\mathfrak{M}$ are orthonormal subsets of H. For events A and B in $\mathfrak{M}$, A oc B if and only if $A \cup B$ is a basis for H; and A op B if and only if there exists event C such that A oc C and B oc C. We need the following two lemmas concerning arbitrary events A, B, C in $\mathfrak{M}$.

4.19. Lemma.

A. A oc C if and only if $C^\perp = \vee A$.
B. A op B if and only if $\vee A = \vee B$.

PROOF. Project. □

4.20. Theorem. *If A and B are events in frame manual $\mathscr{F}(H)$, then*

(i) the following are equivalent:
  (a) A op B,
  (b) $\vee A = \vee B$,
  (c) $A \leftrightarrow B$ (A is logically equivalent to B in manual $\mathscr{F}(H)$);
(ii) $A \perp B$ if and only if $[A] \perp [B]$.

(Be careful about the two meanings of $\perp$ here. The one on the left is orthogonality in the Hilbert space, hence based on the inner product; the one on the right is orthogonality in the logic $\mathbb{L}(H)$, meaning $[A] \leq [B]'$.)

PROOF. Project. □

Theorem 4.20 sums up the main connections between a Hilbert space frame manual and its associated logically structure. Part (i) shows that events A and B are operationally perspective if and only if they are logically equivalent, which happens if and only if they span the same subspace.

Recall that in Equation (3A.1) we defined a weight on the spin manual by means of projections of a unit vector ψ onto subspaces of H. We assumed that you were familiar with the notion of projections in two-dimensional vector spaces. We turn now to a notion of projection in infinite dimensional Hilbert space.

4.21. Definition. If $\mathbb{K} = \{K_1, \ldots\}$ is a countable collection of subspaces in H, then we define

$$\underline{+\mathbb{K}} = \left\{ \sum_{j=1}^{\infty} x_j \mid x_j \in K_j \quad \text{for all} \quad j \in \mathbb{N}, \quad \text{and} \quad \langle\!\langle x_j \rangle\!\rangle \text{ is summable} \right\}.$$

We sometimes write $K_1 + K_2 + \cdots$, or $+_{j\in\mathbb{N}} K_j$ for $+\{K_1, K_2, \ldots\}$.

4.22. Theorem. *If K, L are subspaces in H, and $K \perp L$, then*

(i) *$K + L$ is a subspace in H;*
(ii) *$K + L = K \vee L$.*

PROOF. Project. □

4.23. Theorem (The Finite Projection Theorem). *If K is a subspace in H, then every vector x in H can be written in a unique way as $x = x_1 + x_2$ with $x_1 \in K$ and $x_2 \in K^\perp$.*

PROOF. Project. □

4.24. Theorem. *If $\mathbb{K} = \{K_1, \ldots\}$ is a countable, pairwise orthogonal collection of subspaces in H, then $+\mathbb{K} = \vee \mathbb{K}$.*

PROOF. Project.** □

4.25. Definition. If K is a subspace in H and $x \in H$, then *the projection of x onto K* is the unique vector $x_1 \in K$ such that $x = x_1 + x_2$ with $x_2 \in K^\perp$.

4.26. Theorem (The General Projection Theorem). *If $\mathbb{K} = \{K_1, \ldots\}$ is a countable, pairwise orthogonal collection of subspaces in H, then every vector $x \in \vee \mathbb{K}$ can be written in a unique way as a sum $x = \sum_{j=1}^{\infty} x_j$, where $x_j \in K_j$ for all $j \in \mathbb{N}$. Further, $\|x\|^2 = \sum_{j=1}^{\infty} \|x_j\|^2$.*

PROOF. Project. □

We conclude this chapter with a theorem about compatibility of subspaces in a Hilbert space logic that will provide geometric insight in later chapters.

4.27. Theorem. *Subspaces K and L are compatible in Hilbert space logic $\mathbb{L}(H)$ if and only if*

(i) *$K = (K \wedge L) \vee (K \wedge L^\perp)$, and*
(ii) *$L = (L \wedge K) \vee (L \wedge K^\perp)$.*

PROOF. This is merely an application of Theorem 3B.16 to the logic $\mathbb{L}(H)$. □

CHAPTER 5

Maps on Hilbert Spaces

Part A: Linear Functionals and Function Spaces

As with all mathematical structures, it is important to know what kinds of functions from one Hilbert space to another preserve the important properties of the structure. Besides mathematical interest, the functions on Hilbert space that we study in this chapter provide a crucial link to the notion of an "observable physical quantity" in the Hilbert space formulation of quantum mechanics.

Unless otherwise noted, throughout both parts of this chapter M_1 and M_2 will denote arbitrary inner product spaces, which might not be complete. (We use M to suggest that M is a linear manifold.) To be extremely careful we should distinguish between the structures in M_1 and M_2 by writing $\langle x, y\rangle_1$ and $\langle u, v\rangle_2$, as well as $+_1$ and $+_2$ etc., to draw attention to the fact that we might be dealing with two different inner products and manifold structures. But we shall avoid the notational awkwardness caused by too many subscripts and rely on the reader to avoid confusion by paying close attention to the context.

Also, in this chapter H will, as usual, stand for an arbitrary Hilbert space unless otherwise specified.

5A.1. Definitions. A function $A: M_1 \to M_2$ is called a *linear map from* M_1 *to* M_2 if for all $x, y \in M_1$ and all $\lambda \in \mathfrak{C}$,

(i) $A(x + y) = A(x) + A(y)$, and
(ii) $A(\lambda x) = \lambda A(x)$.

A is called a *conjugate linear map* if it satisfies (i) and

(ii)′ $A(\lambda x) = \lambda^* A(x)$ for all $x \in M_1$ and $\lambda \in \mathfrak{C}$.

We denote by $\underline{\mathrm{Lin}(M_1, M_2)}$ the set of all linear maps from M_1 to M_2.

5A.2. Definitions.

A. If $A \in \mathrm{Lin}(M_1, M_2)$, then we say A is *bounded* if there exists a real number s such that

$$\| A(x) \| \leqq s \| x \| \quad \text{for all } x \in M_1.$$

B. If $A \in \mathrm{Lin}(M_1, M_2)$ is bounded, we define

$$\| A \| = \mathrm{glb}\{s \mid s \in \mathfrak{R}, \text{and } \| A(x) \| \leqq s \| x \| \text{ for all } x \in M_1\}.$$

We call $\| A \|$ the *norm of map A*.

We write $\mathbb{B}(M_1, M_2)$ for the set of all bounded linear maps from M_1 to M_2.

From now on we adopt the convenient notational convention of writing Ax for $A(x)$, except occasionally when we think parentheses are needed for clarity.

5A.3. Lemmas.

A. *If $A \in \mathrm{Lin}(M_1, M_2)$, then $A(0) = 0$.*

For parts B and C suppose $A \in \mathbb{B}(M_1, M_2)$.

B. *If $t \in \mathfrak{R}$ and $\| Ax \| \leqq t \| x \|$ for all $x \in M_1$, then $\| A \| \leqq t$.*
C. *$\| A \| \geqq 0$, and equality holds if and only if $Ax = 0$ for all $x \in M_1$.*

PROOF. Project □

The following definition is useful when talking about bounded linear maps.

5A.4. Definition. For $r \in \mathfrak{R}$ with $r > 0$, A *ball of radius r about the origin* in linear product space M is defined by

$$B_M^r = \{x \in M \mid \| x \| \leqq r\}.$$

5A.5. Theorem. *Suppose $A \in \mathrm{Lin}(M_1, M_2)$. The following are equivalent.*

(i) *A is bounded.*
(ii) *There exists $r \in \mathfrak{R}$ with $r > 0$ such that $A[B_{M_1}^r]$ is a bounded subset of M_2.*
(iii) *For every $r \in \mathfrak{R}$ with $r > 0$, $A[B_{M_1}^r]$ is a bounded subset of M_2.*
(iv) *There does not exist a sequence $\langle\langle x_k \rangle\rangle$ of unit vectors in M_1 with $\lim_{k \to \infty} \| Ax_k \| = \infty$.*

PROOF. Project.* □

5A.6. Examples and Projects.

A. Suppose $n, m \in \mathbb{N}$. Show that $\mathrm{Lin}(\mathbb{C}^n, \mathbb{C}^m) = \mathbb{B}(\mathbb{C}^n, \mathbb{C}^m)$. In other words, all linear maps from $\mathbb{C}^n$ to $\mathbb{C}^m$ are necessarily bounded.

B. Let H be a Hilbert space and K a proper subspace in H. For every $x \in H$, define $P_K(x)$ as the unique vector $x_1 \in K$ such that $x = x_1 + x_2$ with $x_2 \in K^{\perp}$. (See Theorem 4.23.) Show that $P_K \in \mathbb{B}(H, H)$. We call P_K the *projection map of H onto K*.

C. Consider interval $[a, b] \subseteq \mathfrak{R}$. Let H be the Hilbert space $\mathscr{L}_2(a, b)$. (See Example 2.3C.) For every $\psi \in \mathscr{L}_2(a, b)$ define ψ' by

$$\psi'(t) = \begin{cases} \text{the derivative of } \psi \text{ at } t, & \text{if that derivative exists,} \\ 0 & \text{otherwise.} \end{cases}$$

Now define a linear manifold in $\mathscr{L}_2(a, b)$ by

$$M = \{\psi \in H \mid \text{the set of points in } (a, b) \text{ at which } \psi \text{ is not differentiable has Lebesgue measure zero, } \psi' \in H, \text{ and } \psi(a) = \psi(b) = 0\}.$$

Define map $D: M \to H$ by $D\psi = i\psi'$. Finally, for each $n \in \mathbb{N}$ with $n \geqq 2/(b - a)$, define function $\psi_n: \mathfrak{R} \to \mathfrak{R}$ by

$$\psi_n(t) = \begin{cases} n(t - a) & \text{if } a \leqq t \leqq a + 1/n, \\ 2 - n(t - a) & \text{if } a + 1/n \leqq t \leqq a + 2/n, \\ 0 & \text{if } a + 2/n \leqq t \leqq b. \end{cases}$$

Show that for all $n \in \mathbb{N}$ with $n > 2/(b - a)$, $\| \psi_n \|^2 \leqq 2/n$, while $\| D\psi_n \|^2 = 2n$. Thus, $\langle\langle \psi_n \rangle\rangle$ converges in norm to 0, while $\lim_{n \to \infty} \| D\psi_n \| = \infty$. Hence, D is an unbounded linear map.

D. Let $[a, b]$ and $H = \mathscr{L}_2(a, b)$ be as in part C above. For each $\psi \in H$, define $(Q\psi)(t) = t\psi(t)$ for all $t \in (a, b)$. Let

$$M = \{\psi \in H \mid Q\psi \in H\}.$$

Then M is a linear manifold in $\mathscr{L}_2(a, b)$, and $Q: M \to H$ is a linear map. Show that Q is bounded.

The maps in Examples 5A.6C and 5A.6D play important roles in physics. We shall see them again in Chapter 8. (See equations (8B.10) and (8B.12).)

5A.7. Definition. If $A \in \mathrm{Lin}(M_1, M_2)$, then we say A is *continuous* if for every sequence $\langle\langle x_n \rangle\rangle$ in M_1, if $\langle\langle x_n \rangle\rangle$ converges in norm to $x \in M_1$, then $\langle\langle Ax_n \rangle\rangle$ converges in norm to Ax.

5A.8. Theorem. *If $A \in \mathrm{Lin}(M_1, M_2)$, then A is bounded if and only if it is continuous.*

PROOF. Project. □

With every Hilbert space there are associated two important collections of linear maps: the collections Lin(H, H) and Lin$(H, \mathfrak{C})$ ($\mathfrak{C}$ is the set of complex numbers organized as a Hilbert space as in Example 2.3A with $n = 1$). We study these collections next.

5A.9. Definition. If M is an inner product space, the continuous members of Lin$(M, \mathfrak{C})$ are called *continuous linear functionals on* M. We write $\hat{M}$ for the set of continuous linear functionals on M.

We usually use lowercase letters to denote the members of $\hat{M}$, except in the following example.

5A.10. Example. Consider interval $[a, b] \subseteq \mathfrak{R}$, and let $C[a, b]$ be the linear manifold in $\mathscr{L}_2(a, b)$ defined in Example 4.7B. Let $I: C[a, b] \to \mathfrak{C}$ be the integral function defined for all $\psi \in C[a, b]$ by $I(\psi) = \int_a^b \psi \, d\mu$, where μ is Lebesgue measure. Then $I \in \widehat{C[a, b]}$, as can be established by Theorem 1B.6C.

We call the following theorem the Riesz representation theorem after F. Riesz, although there are at least two other theorems in mathematical analysis also called the Riesz representation theorem by other authors. The reader should be careful when searching the literature.

5A.11. Theorem (The Riesz Representation Theorem). *If H is a Hilbert space, and $f \in \hat{H}$, then there exists a unique vector $z \in H$ such that for all $x \in H$, $f(x) = \langle x, z \rangle$. For this z, $\| f \| = \| z \|$.*

(The reader should pay close attention to the two different meanings of the notation $\| \ \|$ in the last sentence of the statement. On the left of the equal sign is the norm of a linear functional; on the right is the norm of a vector.)

PROOF. Project.** □

If H is a Hilbert space, and if we associate with every $z \in H$ the map $\hat{z}$ defined by $\hat{z}(x) = \langle x, z \rangle$ for all $x \in H$, then the Riesz representation theorem assures us that the map $z \to \hat{z}$ is a bijection from H to $\hat{H}$. If we define pointwise addition and scalar multiplication on the set of functions $\hat{H}$, we organize $\hat{H}$ into a vector space, and the map $z \to \hat{z}$ is a linear map. Moreover, we can copy the inner product from H to $\hat{H}$ by defining for all $\hat{z}_1, \hat{z}_2 \in \hat{H}$, $\langle \hat{z}_1, \hat{z}_2 \rangle = \langle z_1, z_2 \rangle$. This provides us with an inner product space $\hat{H}$. In fact, it is a Hilbert space, usually associated with H as follows.

5A.12. Definition. The *dual space* of H is the Hilbert space $\hat{H}$ defined in the preceding remark.

We turn next to the space of bounded linear operators on H.

5A.13. Definition. We call the members of $\mathbb{B}(H,H)$ *bounded linear operators* on H, and we write $\underline{\mathbb{B}(H)}$ for $\mathbb{B}(H,H)$.

5A.14. Theorem. *If $A \in \mathbb{B}(H)$, there is a unique operator $A^{\#} \in \mathbb{B}(H)$ such that for all $x, y \in H$, $\langle x, Ay \rangle = \langle A^{\#}x, y \rangle$. We call $A^{\#}$ the adjoint of A.*

PROOF. Project.* □

5A.15. Theorem. *If A_1, $A_2 \in \mathbb{B}(H)$, then the following are satisfied.*

(i) $\langle x, A^{\#}y \rangle = \langle Ax, y \rangle$.
(ii) $(A_1A_2)^{\#} = A_2^{\#}A_1^{\#}$. *(Juxtaposition means function composition.)*
(iii) *For all* $\lambda \in \mathfrak{C}$, $(\lambda A)^{\#} = \lambda^* A^{\#}$.
(iv) $(A_1 + A_2)^{\#} = A_1^{\#} + A_2^{\#}$.
(v) $A^{\#\#} = A$.

PROOF. Project. □

5A.16. Definition. If $A \in \mathbb{B}(H)$, we call A *Hermitian* (or *self-adjoint*) if $A^{\#} = A$. Equivalently, A is Hermitian if and only if

$$\langle Ax, y \rangle = \langle x, Ay \rangle \quad \text{for all } x, y \in H.$$

5A.17. Examples and Projects. Show that the operators in Examples 5A.6B, 5A.6C, and 5A.6D are all Hermitian.

We now state a technical lemma that we shall cite in later work. Its proof can be found in many introductory books on finite dimensional vector spaces.

5A.18. Lemma. *Suppose n is a positive integer and $A \in \mathrm{Lin}(\mathfrak{C}^n, \mathfrak{C}^n)$.*

A. *A matrix representation of $A^{\#}$ can be found by taking the transpose of the complex conjugate of the matrix representation of A with respect to the standard basis for $\mathfrak{C}^n$.*
B. *If A is Hermitian, its matrix representation with respect to the standard basis for $\mathfrak{C}^n$ is symmetric, and its diagonal entries are real.*

The key connection between the physical universe and the mathematical abstraction known as Hilbert space is provided by Hermitian operators. There are primarily two links in this connection. First, as we shall see in Chapter 7, every Hermitian operator can be associated with a set of real numbers known as the "spectrum" of the operator. These are the numbers that form the set of possible outcomes for physical experiments. Second, every Hermitian operator can be associated with a set of subspaces in the Hilbert space on which it operates, and the compatibility of events in physical experiments is thereby linked to the compatibility of subspaces.

The goal of the Hilbert space model for a physical system, therefore, is to associate a Hermitian operator with every physical quantity for which the system can be measured. These quantities are called "observables." The spectrum of a Hermitian operator is the set of all values possible to obtain by measuring for the observable, and the compatibility of the subspaces associated with Hermitian operators reflects the possibility (or impossibility) of simultaneously determining exact values for a pair of observables.

A Hilbert space model of a physical system provides much more descriptive value than a catalogue of observables. We shall be able to define a "state" of a system at an instant in time by a probability measure on the subspace lattice, as we did for more general lattices in Chapter 3, and thereby describe the dynamical change in observable values as the states of a system change. We shall classify states as "classical" and "nonclassical" and once more look into the question of epistemic and ontological uncertainties.

There is one important fact that we must bear in mind, however, as we construct this link between Hilbert spaces and the physical universe. While Hilbert spaces are a convenient mathematical structure for formulating our assumptions about physics, they are not *necessary* for that formulation. As we have seen in Chapter 3, using only elementary set theory we can very clearly formulate a framework in which to discuss assumptions about the physical universe. All the notions of outcomes of experiments, uncertainty, and nonclassical logic can be neatly and completely described without the deep mathematics required to understand Hilbert space. The question then arises, of course: why use Hilbert space at all? The answer is complicated.

The first mathematical formulation of the foundation of quantum mechanics using Hilbert space was made in 1930 by John von Neumann. His work was widely embraced for at least two reasons. First, it was beautiful mathematics, bringing clarity to a bewildering array of theories explaining quantum phenomena. Second, it linked some controversial theories, such as the uncertainty principle, with well-known mathematical ideas, such as the noncommutativity of certain pairs of Hilbert space operators. This link was heralded as justification for the physical theory, particularly since the Hilbert space formulation fit so nicely with other aspects of physics that were not so controversial.

Since 1930, however, researchers have demonstrated that the Hilbert space formulation for quantum physics is mathematically quite restrictive. They have shown that Hilbert spaces satisfy many properties that are not motivated by physical considerations, and moreover, the restrictions of the Hilbert space formulation prevent reasonable alternative physical theories, particularly in the complicated business of the coupling of physical systems. Yet, while these mathematical restrictions seem unsupported by physical considerations, they have not been shown to be inconsistent with any physical experiments yet devised. So, in defense of the Hilbert space model, many point to its success in predicting outcomes of experiments performed in the laboratory. The difficulty with this defense is that those predications do not rely on the philosophical foundation of the Hilbert space model, but only on solutions to equations that can be formulated in mathematical frameworks more general than Hilbert spaces.

It is clear why the question of whether Hilbert space is necessary for physics is still open and why there is much research spurred by the debate between the foundationalists who find Hilbert space too restrictive and physically unmotivated and the experimentalists who find it a tried and true predictor of events in the laboratory. In this book we shall continue our development of the foundations of quantum physics, both in the general operationalist framework begun in Chapter 3 and in the more orthodox Hilbert space framework for which we are now laying the groundwork. The hope is that, by studying both, the reader will have a better understanding of each.

Part B: Projection Operators and the Projection Logic

In this part we establish the main connection between physical observables and linear operators. First, we need some mathematical results. Unless otherwise noted, H and $\mathbb{B}(H)$ denote the same things that they did in Part A of this chapter.

5B.1. Theorem. *If* $A_1, A_2, A_3 \in \mathbb{B}(H)$ *and* $\lambda \in \mathfrak{C}$, *then*

(*i*) $A_1A_2,\ \lambda A_1,\ \lambda A_2 \in \mathbb{B}(H)$;
(*ii*) $(A_1 + \lambda A_2)A_3 = A_1A_3 + \lambda A_2A_3$ *and* $A_1(A_2 + \lambda A_3) = A_1A_2 + \lambda A_1A_3$;
(*iii*) $(A_1A_2)A_3 = A_1(A_2A_3)$.

PROOF. (i) Clearly A_1A_2, λA_1, and λA_2 are linear operators on H. It is easy to show that they are bounded.

Properties (ii) and (iii) are immediate consequences of the definition of operator composition. □

It is common to call a structure $(\mathbb{A}, +, \cdot, \circ)$ an *associative algebra* if $(\mathbb{A}, +, \cdot)$ is a vector space and $\circ$ is a binary operation satisfying (i)–(iii) of Theorem 5B.1. Thus, we call $\mathbb{B}(H)$ the *algebra of bounded linear operators on H*.

5B.2. Definition. If K is a subspace in H and $A \in \mathbb{B}(H)$, we say K is *invariant under A* if $A[K] \subseteq K$.

5B.3. Theorem. *If* $A \in \mathbb{B}(H)$ *and K is a subspace in H invariant under A, then* $K^{\perp}$ *is invariant under* $A^{\#}$.

PROOF. Project. □

5B.4. Theorem. *If $A \in \mathbb{B}(H)$ and A is Hermitian, then $\langle Ax, x \rangle$ is a real number for every $x \in H$.*

PROOF. Project. □

5B.5. Theorem. *If $A_1, A_2 \in \mathbb{B}(H)$ and both maps are Hermitian, then $A_1 A_2$ is Hermitian if and only if $A_1 A_2 = A_2 A_1$.*

PROOF. Project. □

5B.6. Definition. A member $P \in \mathbb{B}(H)$ is called a *projection operator on H* if and only if $P = PP^{\#}$. We denote the set of all projection operators on H by $\mathbb{P}(H)$.

5B.7. Theorem. *If $P \in \mathbb{B}(H)$, then $P \in \mathbb{P}(H)$ if and only if*

(i) $P = P^{\#}$, *and*
(ii) $PP = P$.

We write P^2 for PP. [If $P^2 = P$ we call P idempotent.]

PROOF. Project. □

5B.8. Theorem. *Let $P \in \mathbb{P}(H)$, and let $K = \text{image}(P) = \{Px \mid x \in H\}$. Then*

(i) *P is the identity map on K;*
(ii) $K = \{x \in H \mid Px = x\}$;
(iii) *K is a subspace in H;*
(iv) *$I - P \in \mathbb{P}(H)$ (I is the identity map on H) and* $\text{image}(I - P) = K^{\perp}$;
(v) *if $x = x_1 + x_2$ with $x_1 \in K$ and $x_2 \in K^{\perp}$, then $Px = x_1$ and $(I - P)x = x_2$;*
(vi) *if $x \in H$ and $\| Px \| = \| x \|$, then $Px = x$;*
(vii) *if $x \in H$, then $\langle Px, x \rangle = \| Px \|^2$;*
(viii) $\| P \| = 1$.

We call P the projection of H onto K.

PROOF. Project.* □

5B.9. Theorem. *If P, Q are projections of H onto K, L, respectively, then the following are equivalent:*

(i) $K \perp L$;
(ii) $PQ = 0$;
(iii) $QP = 0$;
(iv) $P[L] = \{0\}$;
(v) $Q[K] = \{0\}$.

PROOF. Project.* □

5B.10. Definition. Two projections P and Q are *orthogonal*, denoted $P \perp Q$, if any one (hence all) of the properties in Theorem 5B.9 hold.

Properties (iii) and (v) in Theorem 5B.8 show that the map $P \to \text{image}(P)$ is a bijection between $\mathbb{P}(H)$ and $\mathbb{L}(H)$, the set of all subspaces in H. Moreover, this bijection "preserves" orthogonality because of the way we defined orthogonality for projections. Next we shall define an order relation on $\mathbb{P}(H)$ so that we can identify the set of projections with the subspace logic $\mathbb{L}(H)$. (See Theorem 4.18.) We shall need some theorems about how projections behave with respect to joins of subspaces.

5B.11. Definition. A sequence $\langle\langle A_k \rangle\rangle$ in $\mathbb{B}(H)$ is called *summable* if and only if $\langle\langle \| A_k \| \rangle\rangle$ is a summable sequence of real numbers.

5B.12. Theorem. *If $\langle\langle A_k \rangle\rangle$ is a summable sequence in $\mathbb{B}(H)$, then for all $x \in H$, $\langle\langle A_k x \rangle\rangle$ is a summable sequence in H.*

PROOF. Project. □

5B.13. Definition. If $\langle\langle A_k \rangle\rangle$ is a summable sequence in $\mathbb{B}(H)$, we write $A = \sum_{k=1}^{\infty} A_k$ for the function defined by $Ax = \sum_{k=1}^{\infty} A_k x$ for all $x \in H$.

Using theorems from Chapter 2, it is easy to establish that if $\langle\langle A_k \rangle\rangle$ is a summable sequence in $\mathbb{B}(H)$, and $A = \sum_{k=1}^{\infty} A_k$, then $A \in \text{Lin}(H, H)$. A is also bounded, because if $x \in H$, then $\| Ax \| \leq \sum_{k=1}^{\infty} \| A_k \| \| x \| = \| x \| \sum_{k=1}^{\infty} \| A_k \|$. Thus, $\sum_{k=1}^{\infty} A_k \in \mathbb{B}(H)$.

5B.14. Theorem. *If $\langle\langle A_k \rangle\rangle$ is a summable sequence in $\mathbb{B}(H)$, and $A = \sum_{k=1}^{\infty} A_k$, then for all $B \in \mathbb{B}(H)$, $AB = \sum_{k=1}^{\infty} A_k B$, and $BA = \sum_{k=1}^{\infty} BA_k$.*

PROOF. Project. □

5B.15. Theorem. *If $\langle\langle P_k \rangle\rangle$ is a sequence in $\mathbb{P}(H)$ summable to $P \in \mathbb{B}(H)$, then $P \in \mathbb{P}(H)$ if and only if $P_k \perp P_j$ for all $k, j \in \mathbb{N}$ $(k \neq j)$.*

PROOF. Project.* □

5B.16. Theorem. *If $\langle\langle P_j \rangle\rangle$ is a pairwise orthogonal sequence in $\mathbb{P}(H)$ summable to $P \in \mathbb{P}(H)$, and if for every $j \in \mathbb{N}$, $K_j = \text{image}(P_j)$, then*

$$\text{image}(P) = \underset{j \in \mathbb{N}}{+} K_j = \bigvee_{j \in \mathbb{N}} K_j.$$

PROOF. Project. □

5B.17. Definitions. If $P, Q \in \mathbb{P}(H)$ and $K = \text{image}(P)$ and $L = \text{image}(Q)$, then we define

A. $P < Q$ if and only if $K \subset L$ $(K \neq L)$;
B. $\overline{P \vee Q}$ (the *join* of P and Q) is the projection of H onto $K \vee L$;
C. $\overline{P \wedge Q}$ (the *meet* of P and Q) is the projection of H onto $K \cap L$.

5B.18. Theorems.

A. The map $P \to \text{image}(P)$ from $\mathbb{P}(H)$ to $\mathbb{L}(H)$ is a bijection that preserves order, orthogonality, meet, and join.
B. The structure $(\mathbb{P}(H), \leqq, {}^{\perp})$ is a logic with unit I_H [the identity map on H] and zero Z_H [$Z_H(x) = 0$ for all $x \in H$].
C. If $\langle\langle P_k \rangle\rangle$ is a pairwise orthogonal sequence in $\mathbb{P}(H)$, then

$$\bigvee_{k=1}^{\infty} P_k = \sum_{k=1}^{\infty} P_k.$$

PROOF. Project. □

Since the projection operators on H are members of a logic, we call them propositions. They are statements that at any instant in time can be true or false. But what is the physical meaning of a proposition P? The answer is provided by the spectral theorem, which we shall see in Chapter 7. Here is a preview of those ideas.

As we mentioned at the end of Part 5A, we shall associate physical observables with Hermitian operators and with each such operator a set of real numbers called its spectrum, which will be the set of possible values of the observable. The spectral theorem will associate with each Hermitian operator (hence each observable) a function from the Borel sets in $\mathfrak{R}$ to the projection lattice $\mathbb{P}(H)$. If B is a Borel set and P is the projection operator that observable $\mathcal{O}$ associates with P, then the proposition P stands for the statement: "A measurement of $\mathcal{O}$ will result in a value in B." Whether that proposition is true or false at any given instant will depend on the "state" of the physical system at that instant. The states will assign probability values to the propositions.

We conclude this chapter by connecting compatibility of subspaces with the commutativity of the projections onto those subspaces.

5B.19. Theorem. *If $P, Q \in \mathbb{P}(H)$, then*

(i) $PQ = QP$ if and only if PQ is a projection;
(ii) if PQ is a projection, then

$$\text{image}(PQ) = \text{image}(P) \cap \text{image}(Q).$$

PROOF. Project. □

5B.20. Theorem. *If $P, Q \in \mathbb{P}(H)$ and $K = \text{image}(P)$, $L = \text{image}(Q)$, then $PQ = QP$ if and only if K and L are compatible subspaces in the logic $\mathbb{L}(H)$. Equivalently, $PQ = QP$ if and only if P and Q are compatible in $\mathbb{P}(H)$.*

PROOF. Project. □

In Chapter 3 we defined compatibility of propositions in an operational logic strictly from physical considerations involving simultaneous testability of events. Now we see that two propositions in a Hilbert space projection logic are compatible if and only if they commute. This link between logical compatibility of propositions and commutativity of projection operators has been a powerful factor in the attractiveness of Hilbert space as a model for describing the tenets of quantum physics.

CHAPTER 6

State Space and Gleason's Theorem

Introduction

We begin this chapter by defining the state space of a general logic. We examine the geometric structure of the state space and use it to define the notions of pure states, mixtures of states, and physical properties. Next we define an observable on a logic, allowing us to consider physical experiments whose outcome sets are more general than the finite subsets of $\mathfrak{R}$ we saw in Chapter 1. We shall be guided by Lemma 1B.10, however, when we define the expected value of an observable as the integral of the identity function on $\mathfrak{R}$ with respect to a measure determined by a state on the logic.

In Part B of this chapter we see that every observable on a Hilbert space logic $\mathbb{L}(H)$ can be associated with a unique Hermitian operator on H, and every state in the state space can be associated with a vector in H. This will pave the way for the final link between observables, Hermitian operators, and states, which we establish in Chapter 7.

Part A: The Geometry of State Space

6A.1. Definitions.

A. A logic L is *complete* if and only if for all $D \subseteq L$, lub(D) exists. We denote this least upper bound by $\vee D$ or $\bigvee_{d \in D} d$.
B. A logic L is *σ-complete* if $\vee D$ exists for every countable set $D \subseteq L$.
C. A state s on a logic L is *σ-additive* if and only if for every pairwise orthogonal sequence $\langle\langle p_k \rangle\rangle$ in L, if $\bigvee_{k \in \mathbb{N}} p_k$ exists, then $s(\bigvee_{k \in \mathbb{N}} p_k) = \sum_{k=1}^{\infty} s(p_k)$.

We denote by $\Omega_{\sigma,L}$ the set of all σ-additive states on L. As usual, we write simply Ω_σ if there is no doubt about which logic is meant.

All our logics will be assumed to be complete, hence σ-complete. In particular, we have the following for Hilbert spaces.

6A.2. Lemma. *If H is a Hilbert space, then its projection logic $\mathbb{P}(H)$ is a complete logic.*

PROOF. Project. □

Throughout the rest of this chapter, L will be an arbitrary complete logic unless otherwise noted, and Ω_σ will be its set of states.

Let us denote by $\mathfrak{R}^L$ the set of all functions from L to $\mathfrak{R}$. We can organize $\mathfrak{R}^L$ into a vector space using pointwise addition and scalar multiplication. That is, for $\varphi_1, \varphi_2 \in \mathfrak{R}^L$ and $t \in \mathfrak{R}$, define $(\varphi_1 + \varphi_2)(p) = \varphi_1(p) + \varphi_2(p)$, and $(t\varphi_1)(p) = t\varphi_1(p)$ for all $p \in L$. The vector space $\mathfrak{R}^L$ might not have a countable basis; for example, this will be the case if L is uncountable. Clearly $\Omega_\sigma \subseteq \mathfrak{R}^L$.

6A.3. Definition.

$$M_\sigma = \left\{ \sum_{k=1}^{n} \lambda_k s_k \,|\, n \in \mathbb{N}, \quad \lambda_k \in \mathfrak{R}, \quad s_k \in \Omega_\sigma \text{ for } k = 1, \ldots, n \right\}.$$

We call M_σ the *state space* of L.

Observe that the same proof used in Theorem 4.13(i) can be used to show that M_σ is a linear manifold. It is in fact that the smallest linear manifold in $\mathfrak{R}^L$ containing Ω_σ.

The reader should be warned here that we have used the phrase "state space" in a way somewhat different from the way it is used in the general literature on this subject. To deal with infinite logics it is necessary to consider a special norm on M_σ called a "base norm" and to define the state space as the closure of M_σ with respect to the base norm. The geometry of the logics we shall consider can be fully illustrated without the complication of defining the base norm. however, because our most general examples will use finite logics, and our infinite Hilbert space logics come with their own special norms.

6A.4. Definition. If M is a linear manifold and $C \subseteq M$, then C is *convex* in M if and only if for all $x, y \in C$, if $t \in \mathfrak{R}$ and $0 \leqq t \leqq 1$, then $tx + (1-t)y \in C$.

6A.5 Examples and Projects.

A. Show that every interval $[a, b] \subseteq \mathfrak{R}$ is convex in $\mathfrak{R}$.
B. Show that each quadrant in $\mathfrak{R}^2$ is convex in $\mathfrak{R}^2$.
C. Let H be Hilbert space and $r \in \mathfrak{R}$ with $r > 0$. Show that B_H^r is convex in H.

D. Show that the set $D = \{(x,0) | x \in \Re\} \cup \{(0,y) | y \in \Re\}$ is not convex in $\Re^2$.
E. Show that $M = \{\alpha(1,0,0,1) + \beta(0,1,0,0) | \alpha, \beta \in \Re\}$ is a convex set in $\Re^4$.

6A.6. Theorem. Ω_σ *is convex in* M_σ.

PROOF. Project. □

6A.7. Examples and Project. (Example B contains a project.)

A. Let us consider the logic L pictured in Figure 6A.1a. (As we saw in Example 3B.7A, this is the form of the operational logic of a dichotomy.) Let us order the members of L in the order $(0, p, p', 1)$ and represent a function $\varphi \in \Re^L$ by $\varphi = (t_1, t_2, t_3, t_4) = (\varphi(0), \varphi(p), \varphi(p'), \varphi(1))$.

Our first observation is that every members s of Ω_σ is of the form $s = (0, t, 1 - t, 1)$, where $t = s(p)$. We can, therefore, write $\Omega_\sigma = \{(0, t, t - 1, 1) | t \in \Re \text{ and } 0 \leq t \leq 1\}$. If we suppress both the first and fourth constant coordinates of Ω_σ, we can draw M_σ as a two-dimensional vector

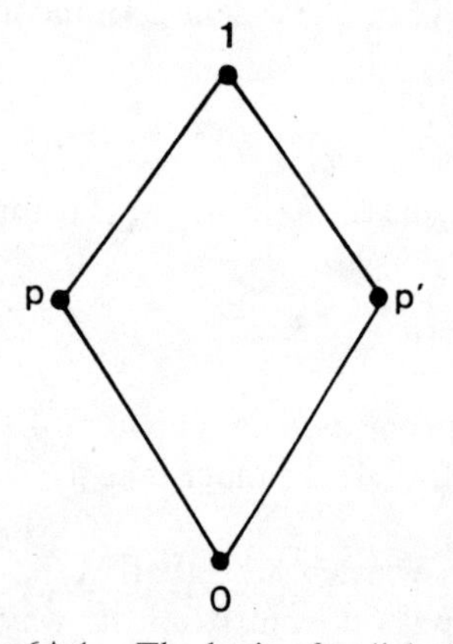

Figure 6A.1a. The logic of a dichotomy.

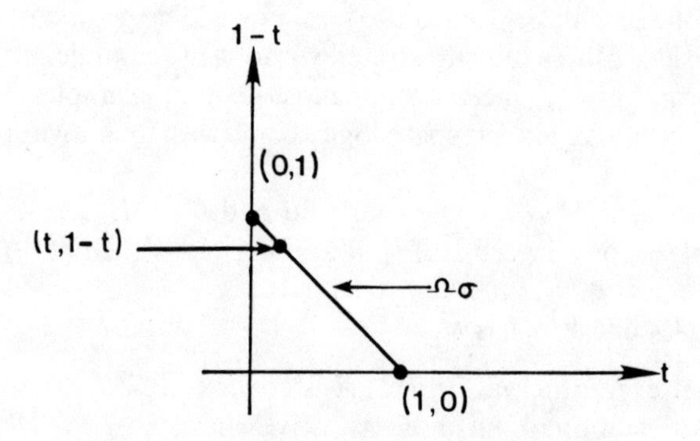

Figure 6A.1b. The state space of dichotomy.

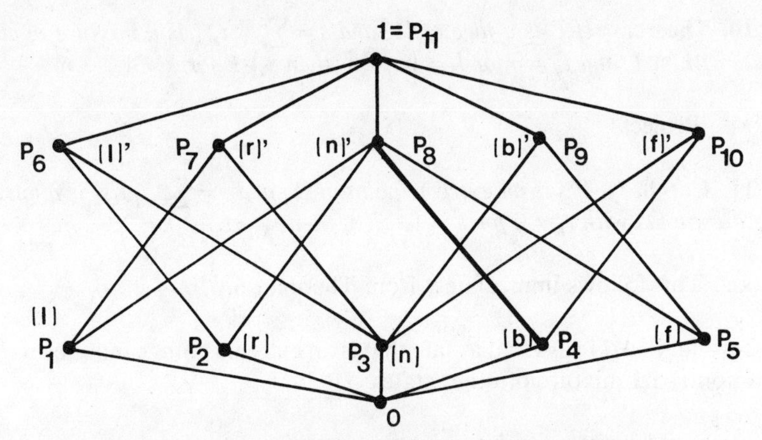

Figure 6A.2. The operational logic for the bowtie manual.

space with Ω_σ as the line segment between $(1, 0)$ and $(0, 1)$, as in Figure 6A.1b. We shall refer to this figure in a future example.

B. Consider the operational logic L for the bow tie manual, as pictured in Figure 6A.2. We can order the members of L, as we did in Example A above, so that $\Omega_\sigma \subseteq \Re^{12}$. Show that if $s = (0, t_1, \ldots, t_{10}, 1)$, then s is uniquely determined by its values $t_2 = s(p_2)$, $t_3 = s(p_3)$, and $t_4 = s(p_4)$. It is important to note that, although a state is determined by only three of its coordinates, the other nine coordinates do not remain constant for all states, so that we *cannot* consider $\Omega_\sigma \subseteq \Re^3$ in imitation of Example A above. Nevertheless, since we shall be concerned primarily with the geometry of Ω_σ, it will be profitable for us to represent Ω_σ by a subset of $\Re^3$, identifying each member s of Ω_σ with the point $(s(p_2), s(p_4), s(p_3))$. Make such a drawing, and compare it with the answer to Project 3A.12.
[CAUTION! Note that the three coordinates of each point representing a state in your drawing are in the order (t_2, t_4, t_3) instead of (t_2, t_3, t_4). This is to make the picture of Ω_σ more recognizable when it is drawn in perspective with $t_2 = s(p_2)$ "coming out of the paper," $t_4 = s(p_4)$ horizontal, and $t_3 = s(p_3)$ vertical.]

6A.8 Definitions.

A. A subset $F \subseteq \Omega_\sigma$ is called a *face* of Ω_σ if and only if $F \neq \emptyset$, F is convex, and for all $x, y \in \Omega_\sigma$, if $t \in \Re$ with $0 < t < 1$, and if $tx + (1 - t)y \in F$, then $x, y \in F$.

B. A state $s \in \Omega_\sigma$ is called an *extreme point* of Ω_σ if and only if $\{s\}$ is a face of Ω_σ.

Think of a face of Ω_σ as a nonempty convex subset that contains no point of any open line segment in Ω_σ unless it contains the entire line segment.

6A.9. Examples and Projects. Find the faces and extreme points of Ω_σ for Examples 6A.7A and 6A.7B.

6A.10. Theorem. *If F is a face of Ω_σ and $s=\sum_{k=1}^n t_k s_k$ is a mixture of states in Ω_σ with $s \in F$ and $t_k \neq 0$ for $k=1,\ldots,n$, then $s_k \in F$ for $k=1,\ldots,n$.*

PROOF. Project.* □

6A.11. Corollary. *If s is an extreme point of Ω_σ and $s=\sum_{k=1}^n t_k s_k$ is a mixture of states in Ω_σ with $t_k \neq 0$ for $k=1,\ldots,n$, then $s_k = s$ for $k=1,\ldots,n$.*

PROOF. This follows immediately from Theorem 6A.10. □

Corollary 6A.11 says that a state that is an extreme point cannot be written as a nontrivial mixture of other states.

Faces and extreme points give us a mathematical way to define two important physical concepts: properties and pure states.

6A.12. Definition. A *property* is a face of Ω_σ.

An example of a property in Ω_σ for the operational logic of the bow tie manual in Example 6A.7B is the face $F=\{(t_2,t_4,t_3)|t_2=1\}$. It represents a property of the system that could be called "right-boldness." To see this, observe that the proposition $p_2=[\{r\}]$ can be interpreted as the proposition "The firefly is lit up on the right side of box." Then

$$F=\{s \in \Omega_\sigma | s(p_2)=1\}.$$

In other words, the property F is exactly the set of states for which the proposition p_2 is true with certainty (probability 1).

Similarly, consider proposition $p_{10}=[\{b,n\}]$, the proposition that the firefly is not lit up in the front of the box. The property

$$G=\{s \in \Omega_\sigma | s(p_{10})=1\}$$

might be called "front-shyness."

The faces F and G lead us to the following definition.

6A.13. Definition. A property F in Ω_σ is called *detectable by a proposition* $p \in L$ if and only if $F=\{s \in \Omega_\sigma | s(p)=1\}$.

6A.14. Example and Project. Find all the detectable properties of the state space of the bow tie manual in Example 6A.7B. List the properties that are not detectable by propositions in L.

6A.15. Definition. A *pure state* is an extreme point of Ω_σ.

We turn now to the physical significance of pure states, as illustrated in the following example.

Suppose we consider an experiment $E_u = \{u\text{-spin up}, u\text{-spin down}\}$, which measures the coordinate of electron spin in a direction u, as in Chapter 3. Then the operational logic for the manual $\mathfrak{M} = \{E_u\}$ is the dichotomy logic discussed in Example 6A.7A, and Ω_σ consists of the line segment shown in Figure 6A.1b. There are two pure states: $s_u = (1,0)$ and $s_d = (0,1)$. The pure state s_u is one in which a measurement for u-spin will yield the outcome u-spin up with certainty (probability 1). Notice that both of these pure states are dispersion-free; that is, they take on only the values 0 and 1.

Now consider the mixture of pure states $s_m = \frac{1}{3}s_u + \frac{2}{3}s_d = (\frac{1}{3}, \frac{2}{3})$. This state is not dispersion-free.

The physical paradigm is:

(1) if many measurements of u-spin are made with the electron always in state s_m, the proportion of measurements that will yield the outcome u-spin up is $\frac{1}{3}$

Another way to put it is:

(2) if the electron is in state s_m, the outcome u-spin up will occur with long-run relative frequency $\frac{1}{3}$.

Still another way:

(3) if the electron is in state s_m, the probability of a measurement yielding outcome u-spin up is $\frac{1}{3}$.

It is very important to recognize the difference between the *epistemic* uncertainty, which the three statements above reflect, and the *ontological* uncertainty, which they would reflect if s_m, a state that is not dispersion-free, were also *pure*, hence not a mixture of other states.

It is the ontological uncertainty of pure states that are not dispersion-free that causes much debate about quantum physics, as we mentioned at the end of Chapter 3. There we saw in the five-chamber firefly system a weight that induced a pure state that was not dispersion-free. We shall explore a simpler version of that system—a three-chamber firefly box—in the next example. Since the three-chamber box involves a quasimanual $\mathcal{Q}$ that is not a manual, we shall be looking at the "weight space" of $\mathcal{Q}$, not the state space of its operational logic, which is not defined if $\mathcal{Q}$ is not a manual. Nevertheless, the weight space provides all the structure we need to see a vivid illustration of the idea of ontological uncertainty.

6A.16. Example and Project. Consider a three-chamber firefly system as in Figure 6A.3a. The quasimanual $\mathcal{Q}$ representing the three experiments corresponding to looking into the three windows is shown in Figure 6A.3b. (You can review Chapter 3 for the connection between the firefly box and quasimanuals.) Even though $\mathcal{Q}$ is not a manual, we can consider the set $W_{\mathcal{Q}}$ of all weights on $\mathcal{Q}$, each of which is determined by a triple of numbers

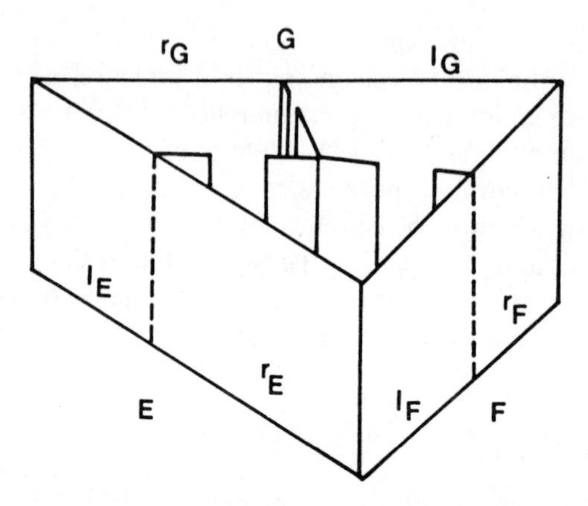

Figure 6A.3a. A box for a three-chamber firefly system.

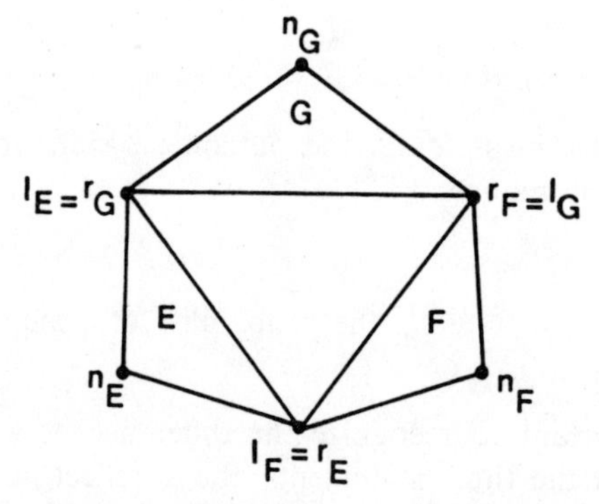

Figure 6A.3b. The quasimanual for a three-chamber firefly system.

(t_1, t_2, t_3) assigned to outcomes r_G $(= l_E)$, r_F $(= l_G)$, and r_E $(= l_F)$, respectively.

Make a sketch of $W_{\mathcal{Q}}$ in $\mathfrak{R}^3$, using the conditions $t_1, t_2, t_3 \in [0, 1]$, $t_1 + t_2 \leqq 1$ (because $r_G \perp r_E$), $t_2 + t_3 \leqq 1$ (because $r_F \perp r_E$), and $t_3 + t_1 \leqq 1$ (because $r_E \perp r_G$). We call $W_{\mathcal{Q}}$ the *weight space* of $\mathcal{Q}$.

If you draw the sketch correctly, you can observe that $\omega = (\frac{1}{2}, \frac{1}{2}, \frac{1}{2})$ is an extreme point of $W_{\mathcal{Q}}$; it is a vertex of a truncated pyramid. Thus, if $\gamma_1, \gamma_2 \in W_{\mathcal{Q}}$ and $t \in \mathfrak{R}$ with $0 < t < 1$ and $\omega = t\gamma_1 + (1-t)\gamma_2$, then $\omega = \gamma_1 = \gamma_2$. In other words, ω cannot be written as a nontrivial mixture of other weights in $W_{\mathcal{Q}}$. So ω is a pure weight that is not dispersion-free. It therefore reflects an ontological uncertainty just like a pure state that is not dispersion-free. The fact that $\omega(r_F) = \frac{1}{2}$ means that the probability of obtaining outcome r_F is $\frac{1}{2}$ if experiment F is performed while the system is in state ω. That $\omega(r_F)$ is not 0 or 1 reflects an uncertainty *in nature*, not an uncertainty about our knowledge of which dispersion-free state the firefly is "really" in.

In Definition 1A.2 we defined the expected value of a finite set of numbers with respect to a weight function. We extend that idea next. For the rest of

Chapter 6A, unless otherwise specified, $\mathbb{B}$ is the collection of Borel sets of real numbers, L is an arbitrary σ-complete logic, Ω_σ is the set of σ-additive states on L, and M_σ is the state space of L.

6A.17. Definitions.

A. An *observable* on L is a function $\mathcal{O}:\mathbb{B}\to L$ such that

(i) $\mathcal{O}(\mathfrak{R})=1_L$;
(ii) if $\langle\!\langle T_j\rangle\!\rangle$ is a pairwise disjoint sequence of Borel subsets of $\mathfrak{R}$, then $\langle\!\langle \mathcal{O}(T_j)\rangle\!\rangle$ is a pairwise orthogonal sequence in L, and $\mathcal{O}(\bigcup_{j=1}^{\infty} T_j)=\bigvee_{j=1}^{\infty}\mathcal{O}(T_j)$.

B. The *resolvent* of an observable $\mathcal{O}$ is defined as

$$\underline{\mathrm{resolv}(\mathcal{O})}=\cup\{U \mid U \text{ is an open set in } \mathfrak{R} \text{ and } \mathcal{O}(U)=0_L\}.$$

C. The *spectrum* of an observable $\mathcal{O}$ is defined as

$$\underline{\sigma(\mathcal{O})}=\mathfrak{R}\backslash\mathrm{resolv}(\mathcal{O}).$$

D. An observable $\mathcal{O}$ is called *bounded* if and only if its spectrum is a bounded subset of $\mathfrak{R}$.

The idea behind an observable is that it represents a physical quantity and its spectrum is the set of possible values of the quantity.

6A.18. Theorems.

A. *If $\mathcal{O}$ is an observable and $s\in\Omega_\sigma$, then $s\circ\mathcal{O}$ (function composition) is a measure for $\mathfrak{R}$ and $s\circ\mathcal{O}(\mathfrak{R})=1$. (The domain of $s\circ\mathcal{O}$ is $\mathbb{B}$, the set of Borel subsets of $\mathfrak{R}$.)*
B. *If $\mathcal{O}$ is a bounded observable and $s\in\Omega_\sigma$, then the identity function I on $\mathfrak{R}$ is integrable over $\mathfrak{R}$ with respect to the measure $s\circ\mathcal{O}$.*

Proof. We leave the proof of A as a project. For B suppose $\sigma(\mathcal{O})$, the spectrum of $\mathcal{O}$, is a subset of interval $[a,b]$. Consider the sequence of simple $s\circ\mathcal{O}$-integrable functions defined by

$$f_n(x)=\begin{cases} a+k\dfrac{b-a}{n} & \text{if } a+k\dfrac{b-a}{n}\leqq x<a+(k+1)\dfrac{b-a}{n} \text{ for } 0\leqq k\leqq n-1,\\ 0 & \text{otherwise.}\end{cases}$$

This sequence converges uniformly to

$$g(x)=\begin{cases} x & \text{if } a<x<b,\\ 0 & \text{otherwise}\end{cases}$$

on the interval (a,b). So g is $s\circ\mathcal{O}$-integrable over $\mathfrak{R}$, and g equals I $s\circ\mathcal{O}$-almost everywhere because $s\circ\mathcal{O}(-\infty,a]=s\circ\mathcal{O}[b,\infty)=0$. This completes the proof of Theorem 6A.18B. □

6A.19. Definition. If $\mathcal{O}$ is a bounded observable, we define a function $\text{Exp}_{\mathcal{O}}:\Omega_\sigma\to\mathfrak{R}$ by

$$\text{Exp}_{\mathcal{O}}(s)=\int_{\mathfrak{R}} I\,d(s\circ\mathcal{O})\quad\text{for every } s\in\Omega_\sigma,\text{ where } I \text{ is the identity function on } \mathfrak{R}.$$

We call $\text{Exp}_{\mathcal{O}}(s)$ the *expected value of $\mathcal{O}$ for state s.*

The physical paradigm for this definition of the expected value is as follows: For observable $\mathcal{O}$ and state s, every Borel set B has measure $s(\mathcal{O}(B))$, which is the probability that a measurement of observable $\mathcal{O}$, made while the physical system is in state s, will yield a value in B. Then the expected value of $\mathcal{O}$ for state s is the "weighted average" of all real numbers, weighted by the measure $s\circ\mathcal{O}$.

Next we view the function $\text{Exp}_{\mathcal{O}}$ as a member of a special set of functions on M_σ.

6A.20. Definition. A function $f:M_\sigma\to\mathfrak{R}$ is called a *bounded linear functional on M_σ* if and only if

(i) $f\in\text{Lin}(M_\sigma,\mathfrak{R})$ (that is, f is a linear map from M_σ to $\mathfrak{R}$), and
(ii) $\{|f(s)|\mid s\in\Omega_\sigma\}$ is a bounded subset of $\mathfrak{R}$.

We denote by M_σ^* the set of bounded linear functionals on M_σ.

6A.21. Theorem. *If $\mathcal{O}$ is a bounded observable, then* $\text{Exp}_{\mathcal{O}}$ *can be extended to a functional* $\overline{\text{Exp}}_{\mathcal{O}}\in M_\sigma^*$ *such that* $\overline{\text{Exp}}_{\mathcal{O}}|\Omega_\sigma=\text{Exp}_{\mathcal{O}}$. (*That is,* $\overline{\text{Exp}}_{\mathcal{O}}(s)=\text{Exp}_{\mathcal{O}}(s)$ *for all* $s\in\Omega_\sigma$.)

PROOF. It follows from Theorem 1B.13 that $\text{Exp}_{\mathcal{O}}(t_1s_1+t_2s_2)=t_1\,\text{Exp}_{\mathcal{O}}(s_1)+t_2\,\text{Exp}_{\mathcal{O}}(s_2)$ for all $s_1,\ s_2\in\Omega_\sigma$ and $t_1,t_2\in\mathfrak{R}$. Then we can extend $\text{Exp}_{\mathcal{O}}$ to a function $\overline{\text{Exp}}_{\mathcal{O}}$ on all of M_σ in an obvious way, because every member of M_σ is a linear combination of members of Ω_σ. The resulting extension is clearly a linear map.

Further, since $\mathcal{O}$ is a bounded observable, there exists $r\in\mathfrak{R}$ with $r>0$ such that the spectrum $\sigma(\mathcal{O})\subseteq[-r,r]$. Thus for all $s\in\Omega_\sigma$ and all $t\in\mathfrak{R}$ with $|t|\leqq 1$, $|\text{Exp}_{\mathcal{O}}(ts)|=|\int_{\mathfrak{R}}I\,d(ts\circ\mathcal{O})|\leqq r|ts(\mathcal{O}([-r,r]))|\leqq r$. This establishes that $\text{Exp}_{\mathcal{O}}[\Omega_\sigma]$ $(=\overline{\text{Exp}}_{\mathcal{O}}[\Omega_\sigma])$ is a bounded set and so completes the proof. □

From now on we write simply $\text{Exp}_{\mathcal{O}}$ to denote the expected value function $\overline{\text{Exp}}_{\mathcal{O}}$ defined on all of M_σ.

6A.22. Definition. A face F of Ω_σ is said to be *exposed by* $f\in M_\sigma^*$ if and only if there exists $t \in \mathfrak{R}$ such that

(i) $\Omega_\sigma \subseteq f^{\leftarrow}(-\infty, t]$, and
(ii) $F = f^{\leftarrow}\{t\} \cap \Omega_\sigma$.

Suppose that $p \in L$ is a proposition. We can define function $f_p: M_\sigma \to \mathfrak{R}$ by $f_p(s) = s(p)$ for all $s \in \Omega_\sigma$. In a manner similar to that used in the proof of Theorem 6A.21 we can extend f_p to a bounded linear functional on all of M_σ. We call the extended functional (which we also denote by f_p) a *propositional functional* on M_σ. We then have that *a face exposed by propositional functional f_p is a property detectable by proposition p.* An important question in current research is: Which members of M_σ^* are propositional functionals? This question has been answered for Hilbert space logics by Andrew Gleason, but it is still open in some more general situations.

The remarkable work of Gleason just mentioned has many manifestations and we shall devote Part B of this chapter to one of them. The main result there will be: for a Hilbert space projection logic $\mathbb{P}(H)$ every member of M_σ^* is associated with a unique Hermitian operator on H. Then in Chapter 7 we shall show that every Hermitian operator on H is associated with a unique observable on $\mathbb{P}(H)$. When we are done we shall have bijections linking all three sets:

$$M_\sigma^* \text{—Hermitian operators on } H\text{—observables on } \mathbb{P}(H).$$

Part B: Gleason's Theorem

We return now to Hilbert space. Throughout this part of Chapter 6 H will stand for an arbitrary Hilbert space, L its projection logic (which we will identify with the subspace logic when convenient), Ω_σ the set of σ-additive states on L, M_σ the state space of L, and M_σ^* the space of bounded linear functionals on M_σ. We begin by defining vector states.

6B.1. Definition. Let x be a nonzero vector in H. We define a function $s_x: L \to \mathfrak{R}$ by

$$s_x(P) = \frac{\|Px\|^2}{\|x\|^2} = \frac{1}{\|x\|^2}\langle Px, x\rangle \quad \text{for every } P \in L.$$

We call s_x a *vector state*.

6B.2. Theorem. *If x is a nonzero vector in H, then $s_x \in \Omega_\sigma$.*

PROOF. Project. □

6B.3. Definition. A vector $y \in H$ is a *mixture* if there exist a pairwise orthogonal sequence $\langle\langle x_k \rangle\rangle$ of unit vectors in H and a sequence $\langle\langle t_k \rangle\rangle$ of real numbers such that

(i) $0 \leqq t_k \leqq 1$ for all $k \in \mathbb{N}$;
(ii) $\sum_{k=1}^{\infty} t_k = 1$;
(iii) $y = \sum_{k=1}^{\infty} t_k x_k$.

6B.4. Theorem. *If $y = \sum_{k=1}^{\infty} t_k x_k$ is a mixture in H, then the function s_y defined for all $P \in L$ by.*

$$s_y(P) = \sum_{k=1}^{\infty} t_k s_{x_k}(P) = \sum_{k=1}^{\infty} t_k \langle P x_k, x_k \rangle$$

is a member of Ω_σ.

PROOF. Project. □

For unit vector x, the state s_x is commonly called a "pure state." Consistent with Definition 3B.22, pure states cannot be written as nontrivial mixtures.

Gleason's theorem asserts that *every* state $s \in \Omega_\sigma$ is a mixture of the form s_y for some $y \in H$. We shall postpone the formal statement of the theorem until the end of the chapter, however, so that we can put it in a larger context.

Next we consider a Hermitian operator on H associated with a set of real numbers in the following way.

6B.5. Definition. If A is a Hermitian operator on H, then a *resolution of A into projection operators* is a pair of sequences: a pairwise orthogonal sequence $\langle\langle P_k \rangle\rangle$ of projection operators on H and a sequence $\langle\langle \lambda_k \rangle\rangle$ of real numbers such that

(i) $A = \sum_{k=1}^{\infty} \lambda_k P_k$, and
(ii) $\sum_{k=1}^{\infty} P_k[H] = H$ or, equivalently, $\sum_{k=1}^{\infty} P_k = \bigvee_{k=1}^{\infty} P_k = 1_L = I_H$, where I_H is the identity operator on H.

We denote the pair of sequences by $\underline{\langle\langle \lambda_k, P_k \rangle\rangle}$.

6B.6. Definition. If x is a nonzero vector in H, A is a Hermitian operator on H, and $\langle\langle \lambda_k, P_k \rangle\rangle$ is a resolution of A into projection operators, then we define the *expected value of A with respect to x* by

$$\underline{\mathrm{Exp}(A, x)} = \sum_{k=1}^{\infty} \lambda_k s_x(P_k). \qquad (*)$$

We justify the term "expected value" as follows. Eventually we shall see an association between a Hermitian operator A and a physical quantity. The association will involve a resolution $\langle\langle \lambda_k, P_k \rangle\rangle$ of A into projection operators, and the physical paradigm will be that $E = \{\lambda_k \mid k \in \mathbb{N}\}$ is an experiment and that for nonzero vector x, s_x induces a weight on E by defining $\omega_x(\lambda_k) = s_x(P_k)$ for all $k \in \mathbb{N}$. With this paradigm it is clear that (*) is just a generalization of the definition of expected value given in Definition 1A.2.

For the remainder of this chapter, unless otherwise noted, A will be an arbitrary Hermitian operator on H, and $\langle\!\langle \lambda_k, P_k \rangle\!\rangle$ will be a resolution of A into projection operators.

6B.7. Theorem. *For nonzero vector* $x \in H$, $\operatorname{Exp}(A, x) = (1/\|x\|^2)\langle Ax, x\rangle$.

PROOF. Project. □

Next we connect A with an observable as defined in 6A.17A.

6B.8. Definition. If A is a Hermitian operator on H with $\langle\!\langle \lambda_k, P_k \rangle\!\rangle$ a resolution of A into projection operators, and $E = \{\lambda_k \mid k \in \mathbb{N}\}$, then define for every Borel set B in the reals

$$\underline{\mathcal{O}_A}(B) = \begin{cases} \bigvee_{\lambda_k \in B} P_k & \text{if } E \cap B \neq \varnothing, \\ 0_L & \text{if } E \cap B = \varnothing. \end{cases}$$

6B.9. Lemma. $\mathcal{O}_A$ *is an observable.*

PROOF. Project. □

6B.10. Theorem. *If x is a nonzero vector in H and $\mathcal{O}_A$ is defined by 6B.8 and is bounded, then*

$$\operatorname{Exp}_{\mathcal{O}_A}(s_x) = \frac{1}{\|x\|^2}\langle Ax, x\rangle.$$

[$\operatorname{Exp}_{\mathcal{O}_A}$ *is defined in 6A.19.*]

PROOF. Project.* □

Theorems 6B.7 and 6B.10 might be considered, respectively, the physical and mathematical versions of the same idea. In Theorem 6B.7 the expected value of A with respect to x is defined (see 6B.6) as the *weighted average* of the experimental outcomes $\{\lambda_k \mid k \in \mathbb{N}\}$ with respect to a weight induced by x, while in Theorem 6B.10 the expected value of $\mathcal{O}_A$ for state s_x is defined (see 6A.10) as the *integral of the identity function on* $\mathfrak{R}$ with respect to a measure determined by x; both expected values are then shown to be equal to $(1/\|x\|^2)\langle Ax, x\rangle$.

Now we are ready to state Gleason's theorem.

6B.11. Theorem (Gleason's Theorem). For every σ-additive state s on the projection logic of a Hilbert space H of dimension at least 3, there is a mixture $y = \sum_{k=1}^{\infty} t_k x_k$ of unit vectors in H such that $s = s_y$. Further, if $f \in M_\sigma^*$, then there exists a unique Hermitian operator $A \in \mathbb{B}(H)$ such that if

$s \in \Omega_\sigma$, and $s = s_y$ for some mixture $y = \sum_{k=1}^{\infty} t_k x_k$ of unit vectors in H, then

$$f(s) = f(s_y) = \sum_{k=1}^{\infty} t_k \langle Ax_k, y_k \rangle,$$

and the sum on the right has the same value for all mixtures y with $s = s_y$.

PROOF. Omitted. This is one of the most significant theorems with one of the most difficult proofs in the entire foundational scheme for orthodox quantum physics.

We learned in Part A of this chapter that for every bounded observable $\mathcal{O}$ for a logic L, the expected value function $\mathrm{Exp}_{\mathcal{O}}$ is a member of M_σ^*. Gleason's theorem paves the way for a converse for Hilbert space projection logics.

To see this, suppose that $f \in M_\sigma^*$. Then we can invoke Gleason's theorem to obtain Hermitian operator A so that every state $s \in M_\sigma$ satisfies

$$f(s) = f(s_y) = \sum_{k=1}^{\infty} t_k \langle Ax_k, x_k \rangle$$

for some mixture $y = \sum_{k=1}^{\infty} t_k x_k \in H$. But, by Theorem 6B.10, if A happens to have a suitable resolution into projection operators, then we have the second equality in the following string:

$$f(s_y) = \sum_{k=1}^{\infty} t_k \langle Ax_k, x_k \rangle = \sum_{k=1}^{\infty} t_k \,\mathrm{Exp}_{\mathcal{O}_A}(s_{x_k}) = \mathrm{Exp}_{\mathcal{O}_A}\left(\sum_{k=1}^{\infty} t_k s_{x_k} \right) = \mathrm{Exp}_{\mathcal{O}_A}(s_y).$$

So we see that $f = \mathrm{Exp}_{\mathcal{O}_A}$. In other words, every member of M_σ^* is an expected value function for some observable.

This last argument works only if A happens to have a suitable resolution into projection operators. Chapter 7 is devoted to the spectral theorem, which shows that every Hermitian operator is indeed suitably associated with a family of projection operators.

CHAPTER 7
Spectrality

Introduction

In the last chapter we saw how Gleason's theorem provided us with out first connection between physical observables and Hermitian operators. Beginning with a Hilbert space H and its projection logic, we defined an observable as a logic-valued function on the real Borel sets. Then through expected values every observable was shown to be associated with a member of $M_\sigma^*(H)$, and then every member of $M_\sigma^*(H)$ was associated with a Hermitian operator on H by Gleason's theorem and the spectral theorem.

Our main result, the spectral theorem, might be called the "fundamental theorem of Hilbert spaces." For every bounded Hermitian operator A, the spectral theorem provides a resolution of A into projection operators and a Borel measure $\mathring{\mu}$ such that for every unit vector x in H the value $\langle Ax, x\rangle$ can be computed as an integral. As we saw in Theorem 6B.10, if x is a unit vector, then $\langle Ax, x\rangle$ is equal to the expected value of an observable on the projection logic of H.

Part A: Finite Dimensional Spaces, the Spectral Resolution Theorem

Throughout Part A of this chapter H is an n-dimensional ($n \geqq 1$) complex Hilbert space, and I_H and 0_H are, respectively, the identity operator and the zero operator on H.

7A.1. Definitions. An *eigenvalue* of a linear operator A on H is a complex number λ such that $A - \lambda I_H$ is not injective. We denote by $\pi(A)$ the set of all eigenvalues of A, and we refer to $\pi(A)$ as *the point spectrum of* A.

7A.2. Lemma. *Let A be a linear operator on H. The following statements are equivalent:*

(i) *λ is an eigenvalue of A.*
(ii) *The value of the determinant* $\det(A - \lambda I_H)$ *is* 0.
(iii) *There is a nonzero vector x in H with $Ax = \lambda x$.*

PROOF. The equivalence of these three ways to say that λ is an eigenvalue of A follows from results found in most textbooks on linear algebra. □

7A.3. Definitions. If λ is an eigenvalue of operator A on H, then the *eigenspace of A* associated with λ is the set

$$S(A, \lambda) = \{x \mid x \in H \text{ and } Ax = \lambda x\}.$$

The nonzero vectors in $S(A, \lambda)$ are called the *eigenvectors* of A associated with λ.

7A.4. Lemmas.

A. *If A is an operator on H, and if λ is an eigenvalue of A, then $S(A, \lambda)$ is a nonzero submanifold of A.*
B. *The eigenvalues of a Hermitian operator are necessarily real numbers.*
C. *If A is a Hermitian operator on H, and λ_1, λ_2 are eigenvalues of A with $\lambda_1 \neq \lambda_2$, then $S(A, \lambda_1) \perp S(A, \lambda_2)$.*

PROOF. Project. □

7A.5. Theorem. *If A is an operator on H, then $\pi(A) \neq \varnothing$.*

PROOF. If A is an operator on H, then $\det(A - \lambda I_H)$ is an n-degree complex polynomial with at least one root. In view of Lemma 7A.2(ii) this completes the proof. □

7A.6. Examples and Projects. Let H be two-dimensional complex Hilbert space. For each of the operators on H that have the following matrix representations, find all eigenvalues and their associated eigenspaces. None of these has more than two eigenvalues.

$$\begin{bmatrix} 1 & 2 \\ -8 & 11 \end{bmatrix}, \quad \begin{bmatrix} 1 & 0 \\ 0 & 1 \end{bmatrix}, \quad \begin{bmatrix} a & b \\ -b & a \end{bmatrix} \quad (a, b \text{ are real numbers, } b \neq 0),$$

$$\begin{bmatrix} 3 & 0 \\ 0 & 3 \end{bmatrix}, \quad \begin{bmatrix} \cos\theta & e^{-i\phi}\sin\theta \\ e^{i\phi}\sin\theta & -\cos\theta \end{bmatrix}.$$

[Hint: Verify that the last matrix has eigenvectors $\begin{bmatrix} e^{-i\phi}\cos\frac{\theta}{2} \\ \sin\frac{\theta}{2} \end{bmatrix}$ and $\begin{bmatrix} -\sin\frac{\theta}{2} \\ e^{i\phi}\cos\frac{\theta}{2} \end{bmatrix}$.]

7A.7. Theorem. *If A is a linear operator on H, then there exist linear subspaces $S_0, S_1, \ldots, S_n$ of H satisfying the following:*

(i) $\{0\} = S_0 \subseteq S_1 \subseteq \cdots \subseteq S_n = H$.
(ii) If k is an integer with $0 \leq k \leq n$, then $\dim(S_k) = k$, and

$$A[S_k] \subseteq S_k \quad \text{and} \quad A[S_k^\perp] \subseteq S_k^\perp.$$

If $A[S_k] \subseteq S_k$, we say that S_k *reduces* A.

Proof. Project.** (Prove this theorem by induction on the dimension of H. The induction is a little tricky.) □

7A.8. Corollary. *If A is a linear operator on H, then there exists an orthonormal basis B_A for H with respect to which the matrix representation of A is a matrix in upper diagonal form:*

$$A = \begin{bmatrix} \lambda_{11} & & ** \\ 0 & \ddots & \lambda_{nn} \end{bmatrix}_{B_A},$$

with the diagonal entries equal to eigenvalues of A and the entries below the diagonal all equal to zero.

Proof. Project. □

7A.9. Theorem. *If A is a Hermitian operator on H, then there exists an orthonormal basis $B_A = \{x_1, \ldots, x_n\}$ for H with respect to which the matrix representation for A is in diagonal form (all nondiagonal entries equal to zero):*

$$A = \begin{bmatrix} \lambda_1 & & 0 \\ 0 & \ddots & \lambda_n \end{bmatrix}_{B_A},$$

and

(i) λ_k is a diagonal entry if and only if λ_k is an eigenvalue of A;
(ii) for each k, $x_k \in S(A, \lambda_k)$;
(iii) for every eigenvalue λ of A,

$$S(A,\lambda) = \mathrm{Span}\,\{x_k \mid x_k \in B_A \text{ and } \lambda = \lambda_k\}.$$

Further,

(iv) if $\pi(A)$ is the set of distinct eigenvalues of A, then

$$H = \bigvee_{\lambda \in \pi(A)} S(A,\lambda).$$

This theorem is sometimes called the "spectral theorem for finite dimensional spaces." We shall reserve that name for another statement that is more readily generalized to infinite dimensional spaces.

PROOF. Let A be a Hermitian operator on H. Let B_A be a basis for H constructed as in Corollary 7A.8. Since A is Hermitian, the upper diagonal matrix representation of A with respect to B_A is symmetric, so it is in the diagonal form desired.

We continue now with the proof of the rest of the theorem.

(i) Since $\det(A - \lambda I_H) = \prod_{k=1}^{n}(\lambda_k - \lambda)$ for every λ in $\mathfrak{C}$, we have that λ is an eigenvalue of A if and only if $\lambda = \lambda_k$ for some $k = 1, \ldots, n$.

(ii) For $1 \leqq k \leqq n$, the representation of x_k with respect to B_A is $\begin{bmatrix} 0 \\ \vdots \\ 1 \\ \vdots \\ 0 \end{bmatrix}$ with 1 in the kth row. It is obvious from this that $Ax_k = \lambda_k x_k$.

(iii) Let λ be an eigenvalue of A, and let $S = \operatorname{span}\{x_k \mid x_k \in B_A \text{ and } \lambda = \lambda_k\}$. For each k with $\lambda = \lambda_k$ we have from (ii) that $x_k \in S(A, \lambda)$. Since $S(A, \lambda)$ is a submanifold of H, we have that $S \subseteq S(A, \lambda)$.

To obtain the reverse inclusion suppose $x \in S(A, \lambda)$. Write x with respect to basis B_A as $x = \sum_{k=1}^{n} \gamma_k x_k$, so that $\lambda x = \sum_{k=1}^{n} \lambda \gamma_k x_k$. Now $\lambda x = Ax = \sum_{k=1}^{n} \gamma_k \lambda_k x_k$. Hence $0 = \sum_{k=1}^{n} \gamma_k (\lambda - \lambda_k) x_k$. Since B_A is linearly independent, we have for each $k = 1, \ldots, n$ either $\lambda = \lambda_k$ or $\gamma_k = 0$. Thus, x is a linear combination of vectors x_k in B_A for which $\lambda = \lambda_k$.

(iv) For each integer k with $1 \leqq k \leqq n$, let Q_k be the one-dimensional subspace of H generated by x_k. From (iii) we have that for each eigenvalue λ of A, $S(A, \lambda) = \bigvee_{\lambda_k = \lambda} Q_k$. Moreover, for distinct eigenvalues λ_1 and λ_2, the spaces $S(A, \lambda_1) = \bigvee_{\lambda_k = \lambda_1} Q_k$ and $S(A, \lambda_2) = \bigvee_{\lambda_k = \lambda_1} Q_k$ are orthogonal by Lemma 7A.4C. Hence $\bigvee_{\lambda \in \pi(A)} S(A, \lambda) = \bigvee_{k=1}^{n} Q_k = \operatorname{span}(B_A) = H$. This completes the proof of the theorem. □

7A.10. Corollary. *If A is a Hermitian operator on H, then $\det(A - \lambda I_H)$ is an n-degree complex polynomial whose roots are the real eigenvalues of A, and the multiplicity of eigenvalue λ is equal to the dimension of eigenspace $S(A, \lambda)$.*

PROOF. This follows immediately from Theorem 7A.9. □

We come now to the main theorem of Chapter 7A. This is the result promised at the end of Chapter 6, at least in the finite dimensional case. Namely, every Hermitian operator has a resolution into projection operators.

7A.11. Theorem (The Spectral Resolution Theorem). *If A is a Hermitian operator on H, then there exist distinct real numbers $\lambda_1, \ldots, \lambda_r$ $(1 \leqq r \leqq n)$ and* *a pairwise orthogonal set of nonzero projections* $\{P_1, \ldots, P_r\}$ *on* H *such that*

(*i*) $\sum_{k=1}^{r} P_k = I_H$, *and*
(*ii*) $A = \sum_{k=1}^{r} \lambda_k P_k$.

Conversely, if $\{\lambda_1, \ldots, \lambda_r\}$ *is a set of distinct real numbers,* $\{P_1, \ldots, P_r\}$ *is a pairwise orthogonal set of nonzero projections on* H, *and* (*i*) *and* (*ii*) *hold, then*

(*i'*) $\{\lambda_1, \ldots, \lambda_r\}$ *is the set of distinct eigenvalues of* A, *and*
(*ii'*) *for each* k, P_k *is the projection of* H *onto eigenspace* $S(A, \lambda_k)$.

PROOF. Let A be a Hermitian operator on H with $\{\lambda_1, \ldots, \lambda_r\}$ the set of its distinct eigenvalues. Recall from Lemma 7A.4B that these eigenvalues are real. From Theorem 7A.9(i) we know that $1 \leqq r \leqq n$. For each integer k with $1 \leqq k \leqq r$, let P_k be the projection of H onto $S(A, \lambda_k)$. From Lemma 7A.4C we know that $\{P_1, \ldots, P_r\}$ is a pairwise orthogonal set of nonzero projections. From Theorem 4.24, therefore, we have that $\sum_{k=1}^{r} P_k = \bigvee_{k=1}^{r} P_k$. Then from Theorem 7A.9(iv), $\bigvee_{k=1}^{r} S(A, \lambda_k) = H$, so $\sum_{k=1}^{r} P_k$ is the projection onto H, namely I_H, and (i) is proved.

To prove (ii) let x be any vector in H. From (i) we can write $x = I_H x = \sum_{k=1}^{r} P_k x$. Then, since $P_k x \in S(A, \lambda_k)$ for each k, we have that $AP_k x = \lambda_k P_k x$ for each k. Thus, $Ax = A(\sum_{k=1}^{r} P_k x) = \sum_{k=1}^{r} A(P_k x) = \sum_{k=1}^{r} (\lambda_k P_k)x$. This completes the proof of (ii).

Let us now assume the hypotheses of the converse.

First we show that if k is an integer with $1 \leqq k \leqq r$, then λ_k is an eigenvalue of A. For such k, $P_k \neq O_H$, so there is a nonzero vector x in image(P_k). Then $P_k x = x$, and for $m \neq k$ we have $P_m x = 0$. Thus, from (ii) we have $Ax = \sum_{m=1}^{r} \lambda_m P_m x = \lambda_k P_k x = \lambda_k x$. Hence λ_k is an eigenvalue of A.

On the other hand, suppose that λ is an eigenvalue of A. Then there is a nonzero vector x with $Ax - \lambda x = 0$. From (i) we have $x = \sum_{k=1}^{r} P_k x$, and from (ii) we have $Ax = \sum_{k=1}^{r} \lambda_k P_k x$. Thus, $0 = Ax - \lambda x = \sum_{k=1}^{r} \lambda_k P_k x - \lambda \sum_{k=1}^{r} P_k x = \sum_{k=1}^{r} (\lambda_k - \lambda) P_k x$. From the pairwise orthogonality of the projections P_k, we have that $B = \{P_k x \mid 1 \leqq k \leqq r \text{ and } P_k x \neq 0\}$ is a linearly independent set. Since $x \neq 0$, we know that $B \neq \varnothing$, so we have that $0 = \sum_{P_k x \in B} (\lambda_k - \lambda) P_k x$. From the linear independence of B follows the fact that $\lambda - \lambda_k = 0$ for some k with $1 \leqq k \leqq r$. This proves that all the eigenvalues of A are in $\{\lambda_1, \ldots, \lambda_r\}$.

It remains to show that each P_k is the projection of H onto $S(A, \lambda_k)$. Let $x \in H$, and consider integer k with $1 \leqq k \leqq r$. From (ii) we have $A(P_k x) = \sum_{m=1}^{r} \lambda_m P_m P_k x = \lambda_k P_k^2 x$, the last equality following from the pairwise orthogonality of the projections P_k. Since $P_k^2 = P_k$, we have $A(P_k x) = \lambda_k (P_k x)$, which shows that $P_k x \in S(A, \lambda_k)$. This concludes the proof of the spectral Resolution Theorem. □

7A.12. Definitions. For $\lambda \in \pi(A)$, we denote by $P^{A,\lambda}$ the projection of H onto $S(A, \lambda)$. When the context leaves no ambiguity about which operator A is under consideration, we write $\underline{P^{\lambda}}$ for $P^{A,\lambda}$. The sum $\sum_{\lambda \in \pi(A)} \lambda P^{\lambda}$ is called the *spectral resolution of A*.

Part B: Infinite Dimensional Spaces, the Spectral Theorem

The theorems in the rest of this chapter are based on results we have developed to this point, but most of their proofs are too complicated to be included in this book. We expect that the reader can, nevertheless, appreciate the beauty of the results and their value for the formulation of quantum mechanics outlined in the next chapter.

We shall state the theorems in this section for bounded operators. Most of these theorems are true in various forms for unbounded operators, which are important for some applications of quantum mechanics. But the statements for unbounded operators are complicated by technicalities that are not necessary for understanding the foundation of quantum mechanics that we discuss in Chapter 8, so it suffices for us to use the simpler statements.

For the remainder of this chapter, H will be a Hilbert space of arbitrary (not necessarily finite) dimension, unless otherwise noted.

The point spectrum of an operator A on a finite dimensional Hilbert space H was defined in terms of the injectivity of an associated operator $T = A - \lambda I_H$. If T is an injective operator on a finite dimensional Hilbert space H, then by the rank-nullity theorem, T is necessarily surjective onto H, and $T^{\leftarrow}$ is a bounded operator on H. If T is an injective operator on an infinite dimensional space, however, even if T is bounded, it might not be surjective onto H. The domain of $T^{\leftarrow}$ might be H or a proper subspace of H and in either case might be a linear transformation that is not bounded. So spectrality in infinite dimensional spaces requires a concept more delicate than just the injectivity of a linear operator.

7B.1. Definition. A *spectral value* of a bounded linear operator A on Hilbert space H is a complex number λ such that $A - \lambda I_H$ does not admit an inverse that is a bounded operator on H. We denote by $\sigma(A)$ the set of all spectral values of A, and we refer to $\sigma(A)$ as the *spectrum of A*.

Whether H is finite or infinite dimensional, if A is an operator on H, then its point spectrum $\pi(A)$ is a subset of its spectrum $\sigma(A)$. (Test your understanding of the definitions of $\pi(A)$ and $\sigma(A)$ by making sure you see why this is true.) As we have said above, all operators on finite dimensional spaces are bounded, so it is possible that $\sigma(A) \neq \pi(A)$ only if H is finite dimensional.

The next theorem provides another characterization of $\sigma(A)$.

7B.2. Theorem. *If A is a bounded operator on Hilbert space H, then $\lambda \in \sigma(A)$ if and only if there exists a sequence $\langle\langle x_k \rangle\rangle$ of unit vectors in H with* $\lim_{k \to \infty} \|(A - \lambda I_H)x_k\| = 0$.

PROOF. Project. □

We move toward integrals with the following generalized notion of a measure.

7B.3. Definition. Let $\mathbb{B}$ denote the collection of Borel sets in $\mathfrak{R}$, and let $\mathbb{P}(H)$ denote the projection logic of Hilbert space H. A *spectral measure* over H is a function $M:\mathbb{B}\to\mathbb{P}(H)$ satisfying:

(i) $M(\mathfrak{R})=I_H$ and $M(\varnothing)=0_H$ (I_H is the identity operator on H; 0_H is the zero operator on H);
(ii) for every pairwise disjoint sequence $\langle\langle K_j\rangle\rangle$ in $\mathbb{B}$, $\langle\langle M(K_j)\rangle\rangle$ is pairwise orthogonal in $\mathbb{P}(H)$ and

$$M\left(\bigcup_{j=1}^{\infty} K_j\right)=\sum_{j=1}^{\infty} M(K_j).$$

7B.4. Lemma. *A spectral measure M over Hilbert space H is an observable on the projection logic $\mathbb{P}(H)$.*

PROOF. This follows immediately from Definition 6A.17 of an observable and the connection between joins and sums in $\mathbb{P}(H)$ provided by Theorem 4.24. □

This lemma is at the very heart of the development of quantum logic. Historically, quantum logics and observables on them were *defined* so that they had the properties of spectral measures, because the Hilbert space model for quantum mechanics was the guiding structure for theoreticians trying to find the abstract, physically motivated essence of quantum physics.

The definition of an observable on a quantum logic in Definition 6A.17 had two motivations—one physical, one mathematical. The physical motivation was used in Chapter 6; it was the paradigm of the spectrum of an observable being the allowable values of a physical quantity. The mathematical motivation is the spectral theorem. It connects Hermitian operators with spectral measure, which, as Lemma 7B.4 states, are observables on Hilbert space logics.

We need the following technical lemma about spectral measures.

7B.5. Lemma. *If M is a spectral measure over H, then for $K, L\in\mathbb{B}$,*

(*i*) *If $K\subseteq L$, then*

$$M(L\backslash K)=M(L)-M(K) \quad \text{and} \quad M(K)\leq M(L);$$

(*ii*) $M(K\cup L)+M(K\cap L)=M(K)+M(L)$ *and* $M(K\cap L)=M(K)M(L)$

[in the last equality juxtaposition means function composition of these two projections];

(*iii*) *if $K\cap L=\varnothing$, then $M(K)\perp M(L)$.*

PROOF. Project.** □

In the following definition we connect spectral measures with ordinary complex-valued measures.

7B.6. Definition. If M is a spectral measure over H, and if $x, y \in H$, then we define a complex-valued function $\mu_{M,x,y}: \mathbb{B} \to \mathfrak{C}$ by

$$\mu_{M,x,y}(K) = \langle M(K)x, y \rangle \quad \text{for every } K \text{ in } \mathbb{B}.$$

7B.7. Lemma. *Let M be a spectral measure over H and let $x, y \in H$. Then $\mu_{M,x,y}$ is a complex-valued measure. If $x = y$, then $\mu_{M,x,y}$ is a real-valued measure.*

PROOF. Project. □

Now we are ready to state the spectral theorem.

7B.8. The Spectral Theorem. *If A is a bounded Hermitian operator on Hilbert space H, then there is unique spectral measure M over H such that for all x and y in H,*

$$\langle Ax, y \rangle = \int_{\mathfrak{R}} I_{\mathfrak{R}} \, d\mu_{M,x,y}. \ (I_{\mathfrak{R}} \textit{ is the identity function on } \mathfrak{R}.)$$

If $x = y$, then the measure $\mu_{M,x,y}$ is real-valued and the integral is a real number.

It might not be obvious that the spectral theorem is a generalization of the spectral resolution theorem (7A.11). We shall show that Theorem 7A.11 implies the existence part of the spectral theorem after a word about notation.

Authors use several different notations to write the equality in the spectral theorem. For example, to emphasize the connection between the complex-valued measure $\mu_{M,x,y}$ and the spectral measure M, some authors write $\langle Ax, y \rangle = \int_{\mathfrak{R}} \lambda \, d\langle M_\lambda x, y \rangle$. This notation can be justified by defining a projction-valued function from $\mathfrak{R}$ to $\mathbb{P}(H)$ by $\lambda \to M_\lambda = M(-\infty, \lambda]$. Then for real interval $(a, b]$ we have $\mu_{M,x,y}(a, b] = \langle (M(a, b]x, y \rangle = \langle (M_b - M_a)x, y \rangle$. This emphasizes the dependence of $\mu_{M,x,y}$ on the endpoints of $(a, b]$ and evokes a version of Lebesgue measure called "Lebesgue–Stieltjes measure."

Another example of a way to write the spectral theorem is to deemphasize the role of x and y, since the equality holds for all x and y. Thus we could write

$$A = \int I_{\mathfrak{R}} \, dM \quad \text{or} \quad A = \int \lambda \, dM_\lambda.$$

This notation emphasizes the relationship of the spectral theorem to the spectral resolution theorem.

Finally, we draw attention to the fact that in Definition 6A.19 we defined the expected value of an observable on a quantum logic as an integral, and in Theorem 6B.10 we connected that integral with an inner product of the form $\langle Ax, x \rangle$ for an observable defined on a Hilbert space logic. Now the spectral theorem expresses $\langle Ax, x \rangle$ as an integral by means of a spectral measure. Historically, this provided mathematical motivation for *defining* an observable in imitation of a spectral measure, as we mentioned after Lemma 7B.4.

Our next lemma states that Theorem 7A.11 implies the existence part of Theorem 7B.8 in the case H is finite dimensional. To show uniqueness, even in this finite dimensional case, requires more machinery than we need to develop at this level. To state our next lemma we need to define a particular spectral measure.

Let A be a Hermitian operator on H with point spectrum $\pi(A) = \{\lambda_1, \dots, \lambda_r\}$ and corresponding projections $\{P^{\lambda_1}, \dots, P^{\lambda_r}\}$. Let us define a spectral measure M_A concentrated at the points in $\pi(A) \subseteq \Re$ as follows: For every Borel set K in $\mathbb{B}$, define

$$M_A(K) = \begin{cases} \sum\limits_{\lambda \in \pi(A) \cap K} P^{\lambda} & \text{if } \pi(A) \cap K \neq \varnothing, \\ 0_H & \text{if } \pi(A) \cap K = \varnothing. \end{cases}$$

7B.9. Lemma. *Let H be a finite dimensional Hilbert space and A a Hermitian operator on H. Let M_A be defined as above.*

(*i*) *Then M_A is a spectral measure.*
(*ii*) *Let $x, y \in H$, and define $\mu_{M_A,x,y}$ as in Definition 7B.6. Then*

$$\langle Ax, y \rangle = \sum_{\lambda \in \pi(A)} \lambda \langle P^{\lambda} x, y \rangle = \int_{\Re} I_{\Re} \, d\mu_{M_A,x,y}.$$

PROOF. Project.* □

One of the most important uses we shall have for the spectral theorem is to help us make the connection between the commutativity of a pair of bounded Hermitian operators and the compatibility of the physical observables they represent. We shall make that connection in the next chapter with the aid of the results obtained in the rest of this chapter.

For the rest of this chapter the phrase "measurable function" refers to Lebesgue measure for $\Re$.

To define a function of a linear operator we need the following theorem.

7B.10. Theorem. *If M is a spectral measure over Hilbert space H, and f is a complex-valued bounded measurable function on $\Re$, then there exists a unique linear operator B on H such that for all x, y in H*

$$\langle Bx, y \rangle = \int_{\Re} f \, d\mu_{M,x,y}.$$

PROOF. Project.** □

7B.11. Definition. Let H be a Hilbert space and A a bounded Hermitian operator on H. Let f be a complex-value bounded measurable function on $\mathfrak{R}$. Then we define operator $f(A)$ as the one and only linear operator on H such that for all x, y in H

$$\langle f(A)x, y \rangle = \int_{\mathfrak{R}} f \, d\mu_{M_A,x,y}.$$

7B.12. Theorem. *If H is a finite dimensional Hilbert space and A is a bounded Hermitian operator on H, and if f is a complex-valued bounded measurable function on $\mathfrak{R}$, then*

$$f(A) = \sum_{\lambda \in \pi(A)} f(\lambda) P^{A,\lambda}.$$

(The projections $P^{A,\lambda}$ are defined in 7A.12.)

PROOF. Project.* □

We remark that the hypotheses that f is bounded and measurable in Theorem 7B.12 are required only because those are the functions for which we have defined $f(A)$. There is no other use we have for these hypotheses in our proof of the theorem. If we were considering only finite dimensional spaces in this chapter, we would have *defined* $f(A)$ by means of the equation in Theorem 7B.12 for every complex-valued function f whose domain contained $\pi(A)$. Our more general definition, however, allows us to state the following important theorem for spaces of arbitrary dimension.

7B.13. Theorem. *Let A and B be bounded Hermitian operators on Hilbert space H. Then a necessary and sufficient condition that $AB = BA$ is that there exist a third bounded Hermitian operator C on H and two complex-valued functions f_A and f_B both defined on $\sigma(C)$ such that*

$$A = f_A(C) \quad \text{and} \quad B = f_B(C).$$

This theorem, proved in 1930 by John von Neumann, was a monumental achievement in the Hilbert space formulation of quantum mechanics. We shall take up the proof of this theorem in the case H is finite dimensional. The proof rests on the following lemmas, which are interesting in their own right.

7B.14. Lemma. *If A is a Hermitian operator on finite dimensional Hilbert space H, then for every λ in $\pi(A)$ there is a polynomial function F_λ with real coefficients such that*

$$P^{A,\lambda} = F_\lambda(A).$$

PROOF. Project.** □

7B.15. Lemma. *If A is a Hermitian operator on finite dimensional Hilbert space H, then an operator B on H commutes with A if and only if B commutes with $P^{A,\lambda}$ for every λ in $\pi(A)$.*

PROOF. If B commutes with $P^{A,\lambda}$ for every λ in $\pi(A)$, clearly B commutes with $A = \sum_{\lambda \in \pi(A)} \lambda P^{A,\lambda}$. Conversely, if B commutes with A, then it commutes with every polynomial $F_\lambda(A)$. By Lemma 7B.14, then, B commutes with $P^{A,\lambda}$ for every λ in $\pi(A)$. This completes the proof. □

7B.16. Corollary. *If A and B are Hermitian operators on finite dimensional Hilbert space H, then A and B commute if and only if $P^{A,\lambda}$ and $P^{B,\gamma}$ commute for all λ in $\pi(A)$ and all γ in $\pi(B)$.*

PROOF. If B commutes with A, it commutes with $P^{A,\lambda}$ for every λ in $\pi(A)$ by Lemma 7B.15. To each $P^{A,\lambda}$, then, we can apply Lemma 7B.15 to conclude that $P^{A,\lambda}$ commutes with $P^{B,\gamma}$ for every γ in $\pi(B)$.

The converse follows easily if we write A and B in their spectral resolutions. □

Now we are ready to prove Theorem 7B.13 in the case H is finite dimensional.

To prove the sufficiency, suppose

$$A = \sum_{\lambda \in \pi(C)} f_A(\lambda) P^{C,\lambda} \quad \text{and} \quad B = \sum_{\lambda \in \pi(C)} f_B(\lambda) P^{C,\lambda}.$$

Then B commutes with itself; hence by Lemma 7B.15 B commutes with each $P^{C,\lambda}$. By Lemma 7B.15 again, therefore, B commutes with A.

To prove the necessity, suppose $AB = BA$. There exists a real-valued function h of two complex variables that is injective on $\pi(A) \times \pi(B)$. Choose such an h. Now define

$$C = \sum_{\substack{\lambda \in \pi(A) \\ \gamma \in \pi(B)}} h(\lambda, \gamma) P^{A,\lambda} P^{B,\gamma}.$$

Next let f_A and f_B be real-valued functions of your choice on $\mathfrak{R}$ such that $f_A(h(\lambda, \gamma)) = \lambda$ and $f_B(h(\lambda, \gamma)) = \gamma$ for all (λ, γ) in $\pi(A) \times \pi(B)$. The proof is completed by the following lemma.

7B.17. Lemma. *Let A, B, C, h, f_A, and f_B be as in the preceding paragraph. Then C is a Hermitian operator and f_A and f_B meet the requirements to complete the proof of Theorem 7B.13 in this case that H is finite dimensional.*

PROOF. Project. □

Let us conclude this chapter with a summary of the different notions of compatibility we have seen so far, for it is this idea that is at the heart of our story of nonclassical physics.

In Definition 3A.8 we called two events in a manual compatible if they could be tested by (were subsets of) a single experiment. In Definition 3B.14 we called two propositions in a logic compatible if they could be supported by a compatibility decomposition. And in Theorem 3B.17 we showed that in an *operational* logic $\Pi(\mathfrak{M})$ compatibility of propositions $[A]$ and $[B]$ is implied by compatibility of representative events A and B in $\mathfrak{M}$.

In Theorem 5B.20 we proved that for the logic of projection operators on a Hilbert space two propositions (projections) are compatible in the logic if and only if they are commuting operators. Finally, in Corollary 7B.16 we showed that two Hermitian operators A and B on a finite dimensional Hilbert space commute if and only if the set of all projections contained in their spectral resolutions forms a pairwise commutative set, which happens if and only if there is a third operator C that permits the observation of A and B simultaneously. This leads us naturally to one last notion of compatibility.

7B.18. Definition. Two Hermitian operators on a Hilbert space H are called *compatible* if and only if they commute.

This brings us to the conclusion of our presentation of the logic of nonclassical physics. We have defined logics in two ways: by starting with a manual, a mathematical concept of a system of laboratory experiments; and by starting with a Hilbert space, the mathematical structure evolving from the ideas of pioneers in physics working in the late 1920s. Both ways share the essential feature of being able to distinguish between a classical logic in which all propositions are compatible and a nonclassical logic that contains incompatible propositions.

In the next chapter we investigate a philosophical dilemma arising from the use of a nonclassical logic to describe physical "reality." It was this dilemma that caused one of the greatest physicists of all time, Albert Einstein, to label quantum mechanics an "incomplete theory."

CHAPTER 8

The Hilbert Space Model for Quantum Mechanics and the EPR Dilemma

Introduction

Throughout this book we have used the phrase "quantum physics" to mean the collection of all the philosophical viewpoints and all the alternative ways of dealing with a physical universe behaving in contradiction to classical Newtonian physics, whereas "quantum mechanics" means a specific set of working hypotheses about the physical universe. Up to now we have been exploring quantum physics in general. Part A of this chapter is a brief history of quantum mechanics. In Part B we outline a set of working hypotheses for quantum mechanics. In Part C we discuss a philosophical dilemma arising from these hypotheses and articulated in a famous paper written in 1935 by Einstein, Podolsky, and Rosen.

Part A: A Brief History of Quantum Mechanics

Many physicists cite December 14, 1900 as the date of the birth of quantum physics. That was the date on which Max Planck delivered to the Physikalischen Gesellschaft in Berlin a paper that proposed a revolutionary theory of energy to explain the relationship between the temperature of a glowing furnace (a "blackbody") and the frequencies of the glowing light. He hypothesized that when the furnace was glowing with a monochromatic light at a specific frequency, then the total energy in the furnace was the sum of a finite number of discrete values of energy. Planck visualized the light in the furnace being produced by a finite number of linear oscillators of some sort, and one consequence of his hypothesis was that each oscillator vibrating at a specific frequency could contribute an energy value only from a discrete set of possibilities. This idea was startling to the world of physics, because classical linear oscillators had been studied very thoroughly, and it was well established that an oscillator at a specific frequency could in general have energies from a continuum of possibilities. Planck was driven to his "desperate" hypothesis because every explanation of blackbody radiation based on classical rules about oscillators had led to relationships between temperature and energy that were simply not validated by experiments.

Planck's departure from classical physics was roundly criticized, but it was not ignored because it provided an experimentally verifiable theory of blackbody radiation that no other theory could produce. For the next 25 years physics was in turmoil. Scientists imitated Planck's approach of dividing energy into discrete quantities, which came to be called "energy quanta," and began developing "quantum theories" of specific heat, electromagnetic radiation, electron energy, and other physical phenomena. The reason for the turmoil was that while these theories were producing remarkably successful predictions verifiable by experiment, they appeared totally ad hoc, without satisfying mechanical models and without the elegant consistency earmarking all of Newtonian physics, which is founded on Newton's simple laws of motion.

In 1924 Louis de Broglie introduced one more measure of chaos into physics. He suggested that the time-honored dichotomy between the physics of particles and the physics of waves be abolished to produce a single theory of "matter-waves" that could behave either like waves or like particles, depending on the experiments performed on them. His idea was an extension of Einstein's theory of light photons. The importance of de Broglie's work was that it predicted the wavelike nature of a stream of electrons, and it provided the philosophical springboard for the invention of quantum mechanics.

Part B: A Hilbert Space Model for Quantum Mechanics

The potpourri of ad hoc quantum assumptions was organized into a single mathematical formulation known as "quantum mechanics" by Erwin Schroedinger and Werner Heisenberg, who independently developed two mathematically equivalent versions in 1925 and 1926. We shall look at Schroedinger's version, known as "wave mechanics," and we shall call this formulation "orthodox quantum mechanics." Space, time, and prudence will restrict our discussion to a most rudimentary presentation, requiring of the reader considerable faith that the hypotheses could be made more plausible with a more thorough discussion. Bear in mind that our objective is not to justify the quantum mechanics model presented but merely to illustrate its use.

We begin with the hypothesis that nature contains "wave-particles," pieces of matter that obey some of the laws of particle physics and some of the laws of wave physics. An electron is one such object. For simplicity we call these objects just "particles."

Let us suppose that we can isolate one particle and that it is "moving back and forth" on some straight line. Notice how our language shapes our imagination. To say that the particle is moving on a straight line really means that we can set up particle detectors along a straight line and observe the signals they send. These signals would be consistent with a model of the particle as a single chunk of mass moving back and forth in accordance with Newtonian particle physics. It is important to emphasize that we are not claiming that we know what the particle *is*, but only what we would *observe* if we set up those particle detectors. We shall return to this subtle point again.

Now suppose that we can attribute to the particle at every instant during its motion a total energy $E = K + V$, where K is its kinetic energy and V its potential energy, the sum E being constant over time. Suppose also that we set up an apparatus with which we can control the potential energy V at each position x along the line on which the particle moves. For example, we can arrange the potential energy along a certain portion of the line so that as the particle moves through that portion its potential energy has values as indicated in Figure 8B.1(a or b). The first is called a "potential well," the second a "potential barrier." Other shapes are also possible.

We now leap to the Schroedinger equation, which is the mathematical equation at the heart of the quantum mechanical description of our moving particle. We give no pretense of a logical development leading to this equation. The literature abounds with heuristic explanations, but every one of them contains at some point a leap of faith based on a blend of classical physics and quantum assumptions. Even Schroedinger abandoned his original rationalization for his equation very soon after he published it. Here we simply state the equation and the recipe for using it to model the "one-particle system" that we described above.

We say that the particle is "described" by a complex-valued function ψ

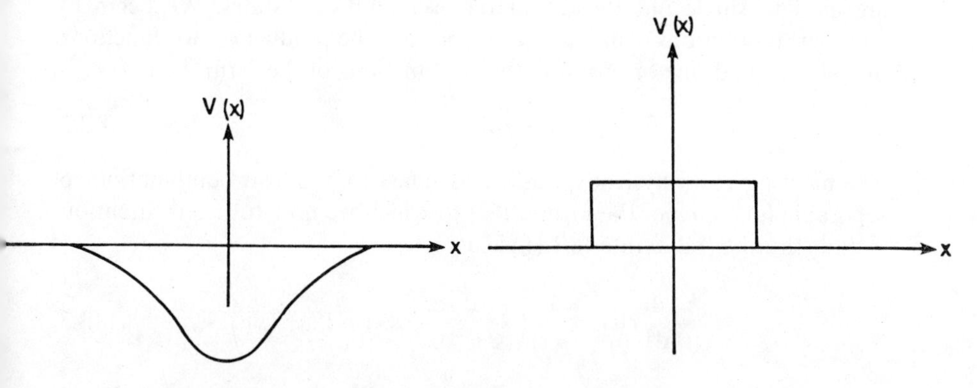

Figure 8B.1. Potential functions for a particle on a line.

of position x (along the line of motion) and time t, where ψ satisfies the equation

$$i\hbar\frac{\partial}{\partial t}\psi(x,t) = -\frac{\hbar^2}{2m}\frac{\partial^2}{\partial x^2}\psi(x,t) + V(x)\psi(x,t), \tag{8B.1}$$

where $\hbar$ and m are constants. This equation is called the *time-dependent Schroedinger equation.*

(In Part C of this chapter we shall consider ψ a function of a third variable to account for "spin," but we need not consider that complication just yet.)

Here is what we mean when we say the function "describes" the particle:

Hypothesis (*). *Each solution ψ of (8B.1) that satisfies certain boundary conditions is called a state of the particle, and the values of the physical quantities associated with the particle are determined by the states. If at time t the particle is associated with state ψ, we say the particle is "in state ψ" at time t.*

Hypothesis (**). *Each state ψ associates the position of the particle with a probability measure M^ψ on the real numbers as follows:*

If the particle is in state ψ at time t, then the probability that a measurement for the position of the particle at time t will yield a value in real interval $[a,b]$ is given by

$$M_t^\psi[a,b] = \int_a^b |\psi(x,t)|^2\,\mathrm{d}x.$$

The above hypotheses, which can be considered a recipe for using equation (8B.1), were not provided by Schroedinger but by Max Born. In fact, the quantum cookbook was written by many authors over several years following 1925, and the condensed version we present here is the result of many years of revision.

Let us now look at the conditions on solutions to equation (8B.1) that are used to single out those that represent physical states. We begin by considering solutions ψ that are separable into the product of two functions, one of time and one of position, that is, functions of the form

$$\psi(x,t) = f(t)\gamma(x). \tag{8B.2}$$

The most general physical states are assumed to be linear combinations of separable functions of the form (8B.2) to which we now turn our attention.

Substituting (8B.2) into (8B.1) yields

$$\frac{i\hbar}{f(t)}\frac{\mathrm{d}}{\mathrm{d}t}f(t) = \frac{1}{\gamma(x)}\left(\frac{-\hbar^2}{2m}\frac{\mathrm{d}^2}{\mathrm{d}x^2}\gamma(x) + V(x)\gamma(x)\right). \tag{8B.3}$$

Since the right side of equation (8B.3) does not depend on t and the left side does not depend on x, both must be equal to a constant, which we call

E. Thus, we have

$$i\hbar \frac{\mathrm{d}}{\mathrm{d}t} f(t) = E f(t) \tag{8B.4a}$$

and

$$\frac{-\hbar^2}{2m} \frac{\mathrm{d}^2}{\mathrm{d}x^2} \gamma(x) + V(x)\gamma(x) = E\gamma(x). \tag{8B.4b}$$

Equation (8B.4b) is called the *time-independent Schroedinger equation* or just the *Schroedinger equation.* We can write equation (8B.4b) another way:

$$\left(-\frac{\hbar^2}{2m} \frac{\mathrm{d}^2}{\mathrm{d}x^2} + V(x)\right)\gamma(x) = E\gamma(x). \tag{8B.5a}$$

Equation (8B.5a) is an eigenvalue equation of the form

$$H\gamma = E\gamma, \tag{8B.5b}$$

where H is an operator on Hilbert space $\mathscr{L}^2(\mathfrak{R})$ and E is an eigenvalue associated with eigenvector γ. (Despite the fact that up to now in this book we have always used H to stand for a Hilbert space, here we use H to denote an operator, because nearly every physics book uses H and calls it the "Hamiltonian operator.")

Continuing our abstraction of equation (8B.5b), if γ is a suitable solution for a fixed value of E, then we call γ a *stationary state*, meaning a state of total energy E constant with respect to time. Then looking at equation (8B.5a), we call $V(x)$ the *potential energy operator*, $(-\hbar^2/2m)(\mathrm{d}^2/\mathrm{d}x^2)$ the *kinetic energy operator*, and their sum the *total energy operator* H. It is also common to call H the *Hamiltonian operator*.

We arrived at the time-independent Schroedinger equation by considering solutions ψ of (8B.1) that are separable into functions of t and x. To qualify as physical states, however, solutions of equation (8B.1) must satisfy a boundary condition, which we consider in the context of the following example.

Let us consider the potential function associated with a particle moving back and forth in the manner of a linear harmonic oscillator. This is given by

$$V(x) = \tfrac{1}{2}m\omega^2 x^2, \tag{8B.6}$$

where m is called the "mass" of the oscillator and ω is its "frequency of vibration." Note the blend of classical and quantum physics. We are dealing with a wave-particle, yet we use the terminology, intuition, and formulas associated with classical particles. Many books on elementary physics explain why equation (8B.6) is the correct potential function for a classical linear harmonic oscillator.

Our task now is to find suitable solutions to equations (8B.4a) and (8B.4b).

We can easily solve the differential equation (8B.4a), and we shall restrict our attention to solutions of the form

$$f(t) = \mathrm{e}^{(-\mathrm{i}Et/\hbar)} \tag{8B.7}$$

so that $|f(t)| = 1$ for all t.

We turn next to solutions of equation (8B.4b), which, for our linear harmonic oscillator, has the form

$$\left(\frac{-\hbar^2}{2m}\frac{d^2}{dx^2}+\tfrac{1}{2}m\omega^2x^2\right)\gamma(x)=E\gamma(x). \tag{8B.8}$$

We seek values for E for which equation (8B.8) has solutions γ satisfying certain boundary conditions. What should those boundary conditions be? According to Hypothesis (**) above, $\psi(x,t)=f(t)\gamma(t)$ is to be interpreted as a probability density in the sense that for fixed t the probability of finding the particle in real interval $[a,b]$ on the x axis is $\int_{\Re}|\gamma(x)|^2\,dx$. (Recall that $|f(t)|=1$ for all t.) Since the probability of finding the particle somewhere on the x axis must be 1, we require that $\int_{\Re}|\gamma(x)|^2\,dx=1$, which implies that

$$\lim_{x\to\pm\infty}|\gamma(x)|=0. \tag{8B.9}$$

It is not a trivial problem to find all the values of E in equation (8B.8) for which there are solutions γ satisfying this boundary condition. Nevertheless, the problem has been solved. (See Merzbacher, for example.) We shall simply state the solution:

> The Schroedinger equation (8B.8) has solutions γ satisfying the boundary condition (8B.9) only for those values E of the form $E=\hbar\omega(n+\frac{1}{2})$, where n is a nonnegative integer.

If we make the convenient substitution $\xi=(\sqrt{m\omega/\hbar})x$, then the solutions to (8B.8) are functions of ξ, the first four of which are shown in Figure 8B.2.

We have reached the central point in this orthodox quantum mechanical description of a one-particle system through the Schroedinger equation. We summarize the model with four statements:

(QM 1) The system is governed by the time-dependent Schroedinger equation whose solutions $\psi(x,t)$ are complex-valued functions of position x and time t.

(QM 2) The solutions are assumed to be linear combinations of functions of the form $\psi(x,t)=f(t)\gamma(x)$, where $|f(t)|=1$ for all t, and γ satisfies the boundary condition $\lim_{x\to\pm\infty}|\gamma(x)|=0$.

(QM 3) A solution ψ is called a *state* of the system, and each state describes the particle in the sense that a measurement for the position of the particle while it is in state ψ will yield a value in interval $[a,b]$ with probability $\int_a^b|\psi(x,t)|^2\,dx=\int_a^b|\gamma(x)|^2\,dx$.

(QM 4) The separability assumption in QM 2 leads to a time-independent Schroedinger equation of the form $H\gamma=E\gamma$, where H is a Hermitian operator on Hilbert space $\mathscr{L}^2(\Re)$. The eigenvalues E for which this equation has solutions satisfying the boundary condition in QM 2 are assumed to be all the possible values of the total energy of the system. An eigenvector γ is called a *stationary state* of the system,

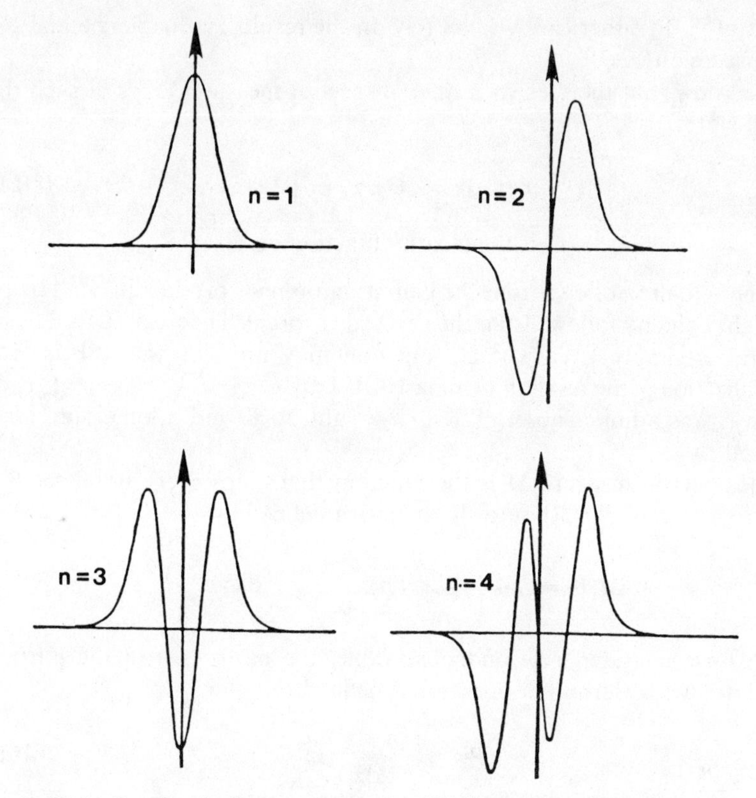

Figure 8B.2. Graphs of solutions to equation (8B.8) corresponding to four values of E.

and while the system is one of these states its total energy is equal to the corresponding eigenvalue.

We complete our discussion of the orthodox quantum mechanical description of a one-particle system by asserting that *all* observables for the system are associated with Hermitian operators on $\mathscr{L}^2(\mathfrak{R})$. (Recall how Gleason's theorem provides the underpinning for this assertion.) Let us consider two observables.

First, consider the Hermitian operator associated with the position of a particle on a straight line. This operator is defined by

$$Q\gamma(x) = x\gamma(x). \tag{8B.10}$$

It can be shown that the spectrum of this observable is the entire real line, as it should be if the spectrum is to be the set of all possible values of the position observable. That Q is thus not a bounded operator presents some technical difficulties, but as we mentioned in Chapter 7, they can be dealt with. They can be avoided altogether if we restrict the motion of the particle to a bounded subset of $\mathfrak{R}$. In either case we can rely on the results in Chapters 6 and 7 in our discussion of Q.

We know that there exists a unique spectral measure M for $\mathfrak{R}$ such that for all states $\gamma \in \mathscr{L}^2(\mathfrak{R})$

$$\int_{\mathfrak{R}} Q\gamma(x)\gamma(x)\,\mathrm{d}x = \langle Q\gamma, \gamma\rangle = \int_{\mathfrak{R}} I_{\mathfrak{R}}\,\mathrm{d}\mu_{M,\gamma,\gamma} \tag{8B.11}$$

($I_{\mathfrak{R}}$ is the identity function on $\mathfrak{R}$).

The left equality follows from the definition of inner product in $\mathscr{L}^2(\mathfrak{R})$, and the right equality follows from the spectral theorem (Theorem 7B.8). The left integral is simply $\int_{\mathfrak{R}} x|\gamma(x)|^2\,\mathrm{d}x$, and you may now see that (8B.11) is a generalization of the result in Lemma 1B.10: $\mathrm{Exp}(E,\omega) = \sum_{x\in E} x\omega(x) = \int_{\mathfrak{R}} I_{\mathfrak{R}}\,\mathrm{d}\mu$, where E was a finite subset of $\mathfrak{R}$, ω a weight on E, and μ a measure for $\mathfrak{R}$ induced by ω.

The spectral measure M is the function that maps every Borel set B to the projection of $\mathscr{L}^2(\mathfrak{R})$ onto the subspace defined by

$$M(B) = \mathrm{clos}\left\{\gamma \in \mathscr{L}^2(\mathfrak{R}) \,\middle|\, \int_{\mathfrak{R}\setminus B} |\gamma|^2\,\mathrm{d}x = 0\right\}.$$

Next we consider a second observable, the *momentum* of the particle, associated with Hermitian operator P defined by

$$P\gamma(x) = \frac{\hbar}{i}\frac{\mathrm{d}}{\mathrm{d}x}\gamma(x). \tag{8B.12}$$

We shall not dwell here on the heuristic arguments leading to the identification of this operator with the physical concept of momentum. (See Merzenbacher, for example.) The observation we wish to make about operators P and Q is that they do not commute, which can be verified by a routine calculation. Hence they are not compatible observables according to Definition 7B.18. Let us restate this fact in the various ways in which we have defined compatibility in previous chapters. Let $\mathbb{B}$ denote the collection of Borel sets in $\mathfrak{R}$.

According to Theorem 7B.13 there does not exist a Hermitian operator C such that $P = f_P(C)$ and $Q = f_Q(C)$. That is, the position and momentum values cannot be written as a function of the spectral values of some third observable.

If M_P and M_Q are the spectral measures associated with P and Q by the Spectral Theorem (7B.8), then

$$\{M_P(B) \mid B \in \mathbb{B}\} \cup \{M_Q(B) \mid B \in \mathbb{B}\}$$

is not a classical (pairwise compatible) set of projections.

Let us look at this second statement more closely. It asserts that there exist Borel sets B_1 and B_2 such that $M_P(B_1)$ is not compatible with $M_Q(B_2)$.

From Theorem 3B.16 we then have that one of the following inequalities is false:

$$M_P(B_1) \leqq (M_P(B_1) \wedge M_Q(B_2)) \vee (M_P(B_1) \wedge M_Q(B_2)'),$$
$$M_Q(B_2) \leqq (M_P(B_1) \wedge M_Q(B_2)) \vee (M_P(B_1)' \wedge M_Q(B_2)). \tag{$*$}$$

In our physical paradigm $M_P(B_1)$ is the proposition: "A measurement of the momentum of the particle will yield a value in Borel set B_1." The proposition $M_Q(B_2)$ is stated analogously. If we suppose, for example, that it is the first of the two inequalities above that is false, we are faced with the following logical relationship between propositions:

We cannot conclude from the fact that a measurement of momentum will yield a value in B_1 that either

(i) a measurement of momentum will yield a value in B_1, and a (simultaneous) measurement of position will yield a value in B_2, or
(ii) a measurement of momentum will yield a value in B_1, and a (simultaneous) measurement of position will not yield a value in B_2.

This logical state of affairs is certainly counterintuitive to our classical understanding of how the world behaves. On the other hand, to insist that both inequalities in (*) must hold for all B_1 and B_2 is to tacitly assume that the momentum and position of the particle can be measured simultaneously for all pairs of Borel sets—in other words, to assume that the position and momentum of the particle can be measured simultaneously to arbitrary degrees of accuracy. This is the assumption that we give up for the sake of preserving the Hilbert space formulation for the position and momentum of this one-particle system.

This concludes our sketch of an example of orthodox quantum mechanics using the Schroedinger equation. As elementary as this example is, it reveals the heart of the logical structure of the Hilbert space formulation. If this discussion has done its job, then it has left the reader with many questions: What are the physical considerations that led Schroedinger to his equation? What is the metaphysical basis for the acceptance of the fundamental probabilism in nature implicit in the interpretation of the solutions to the Schroedinger equation as probability density functions? What experimental evidence is there that lends plausibility to the orthodox formulation of quantum mechanics? What are some of the alternative formulations, and what are their relationships to the orthodox formulation and to current experimental data? What portion of our universe behaves according to the tenets of orthodox quantum mechanics? The literature abounds with volumes addressing these questions, but it is fair to say that the scientific community has not yet been presented with a definitive set of answers acceptable to all. Among the prominent displays of the limitations of orthodox quantum mechanics is the "thought experiment" that we take up in Chapter 8C.

Part C: The EPR Experiment and the Challenge of the Realists

In Part B of this chapter we examined the time-independent Schroedinger equation (8B.4b) for a single wave-particle moving in a straight line. We considered certain solutions of the equation, functions of a single space variable x, and called them stationary states of the single-particle system. As we saw in Chapter 3, however, Stern and Gerlach observed a property of moving particles called "spin" that we have not yet accounted for in our Hilbert space model of orthodox quantum mechanics.

We shall insert spin into our model by expanding our notion of stationary state $\gamma(x)$ to a function $\Gamma(x, s)$ of two variables, position x and spin s. We again consider the simplified situation where Γ is separable into two functions $\Gamma(x, s) = \gamma(x)\chi_z(s)$, where γ is a stationary state and χ_z is a "spin function" or "spin state," which we are about to discuss. Let us emphasize again that the word spin should not be taken literally to mean that our particle is exactly like a marble spinning around some diameter. It is better to think of "z-spin-up" and "z-spin-down" merely as two outcomes in an experiment measuring deflection in a given direction z. We shall actually define a spin state as a solution to an eigenvalue equation, in analogy with equation (8B.5), but first we provide a motivating discussion.

To discuss spin we consider the case that our particle is an electron. The same ideas would apply to other particles, but the numbers would be different. In Figure 8C.1 we illustrate an apparatus that can measure deflection produced by a strong magnetic field in a direction z orthogonal to the direction of the path of the electron. We denote by χ_z a complex-valued function of a discrete variable s that takes only two values: $s = +1$ and $s = -1$. In keeping with our interpretation of $\int_a^b |\gamma(x)|^2\,dx$ as the probability that a measurement for position will yield a value in interval $[a, b]$, we interpret $|\chi_z(+1)|^2$ as the probability that a measurement for the z-spin will yield the outcome z-spin-up. Similarly, $|\chi_z(-1)|^2$ is the probability associated with outcome z-spin-down.

Let us turn to the mathematics that will produce the spin states as eigenvectors of a Hermitian operator, just as we found the states γ as eigenvectors of the Hamiltonian operator. For the time being we consider spin deflection only in the direction z pictured in Figure 8C.1. Later we shall consider spin deflection in other directions.

First we adopt some simplifying notation commonly used by physicists. We denote a spin state χ_z by a column vector $\chi_z = \begin{bmatrix} a \\ b \end{bmatrix}_z$, where $a = \chi_z(+1)$ and $b = \chi_z(-1)$. We then have two special z-spin state functions,

$$\underline{\chi_z^{\text{up}}} = \begin{bmatrix} 1 \\ 0 \end{bmatrix}_z \quad \text{and} \quad \underline{\chi_z^{\text{dn}}} = \begin{bmatrix} 0 \\ 1 \end{bmatrix}_z,$$

which we call the *z-spin-up* and *z-spin-down* states, respectively.

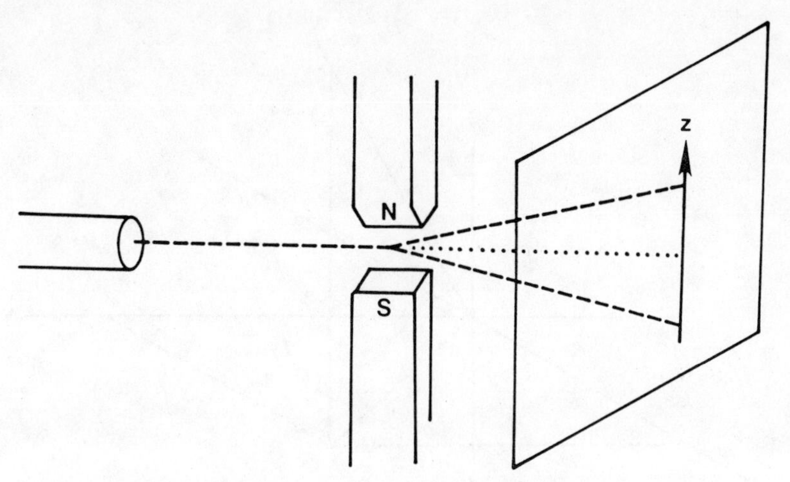

Figure 8C.1. An apparatus to measure electron spin in direction z.

We seek a "Schroedinger equation" for spin. It will be an eigenvalue equation of the form

$$S_z \chi_z = \lambda_z \chi_z \tag{8C.1}$$

where S_z is a 2×2 Hermitian matrix. The eigenvalues associated with a spin state χ_z will be the two possible values of the physical quantity z-spin for an electron. Unlike the physical considerations leading to the Schroedinger energy operator H, it is quite easy to follow the calculations leading to a definition of S_z. (See McWeeny, for example.) We need not trace those calculations for our purposes, however, so we simply write down the Hermitian operator representing the z-spin for an electron in a Stern–Gerlach apparatus:

$$S_z = \frac{1}{2}\begin{bmatrix} 1 & 0 \\ 0 & -1 \end{bmatrix} \tag{8C.2}$$

If we were considering a particle other than an electron, the factor $\frac{1}{2}$ might be a different number. We then define the *z-spin states* as the eigenvectors of the operator S_z.

It is easy to see that $\chi_z^{\mathrm{up}} = \begin{bmatrix} 1 \\ 0 \end{bmatrix}_z$ and $\chi_z^{\mathrm{dn}} = \begin{bmatrix} 0 \\ 1 \end{bmatrix}_z$ are eigenvectors of S_z with eigenvalues $+\frac{1}{2}$ and $-\frac{1}{2}$, respectively. These are the dispersion-free z-spin states, the states for which a measurement for z-spin will result in outcome z-spin-up *with certainty* (because $|\chi_z^{\mathrm{up}}(+1)|^2 = 1$) or outcome z-spin-down *with certainty* (because $|\chi_z^{\mathrm{dn}}(-1)|^2 = 1$).

Now we consider measuring electron spin in directions other than z. Refer to Figure 8C.2. If w is a direction in u-v-z space determined by angle θ between w and the positive z axis and angle φ between the positive v axis and the

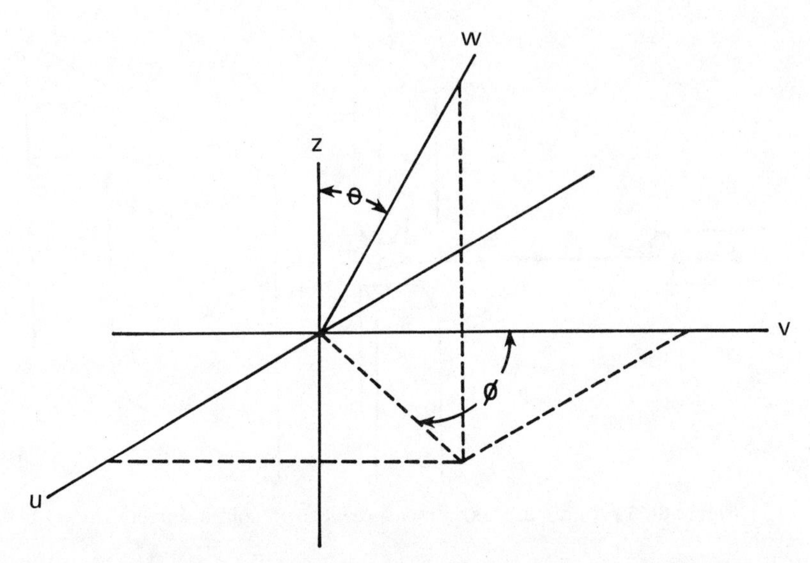

Figure 8C.2. Angles determined by three directions of measurement of electron spin.

projection of w onto the u-v plane, then *w-spin* is an observable represented by matrix

$$S_w = \frac{1}{2}\begin{bmatrix} \cos\theta & e^{-i\varphi}\sin\theta \\ e^{i\varphi}\sin\theta & -\cos\theta \end{bmatrix} \quad \text{where } 0 \leqq \theta \leqq \pi \text{ and } 0 \leqq \varphi \leqq \pi.$$

The dispersion-free eigenstates of S_w are of the form

$$\chi_w^{\mathrm{up}} = \begin{bmatrix} 1 \\ 0 \end{bmatrix}_w = \begin{bmatrix} e^{-i\varphi}\cos(\theta/2) \\ \sin(\theta/2) \end{bmatrix} \quad \text{and} \quad \chi_w^{\mathrm{dn}} = \begin{bmatrix} 0 \\ 1 \end{bmatrix}_w = \begin{bmatrix} -\sin(\theta/2) \\ e^{i\varphi}\cos(\theta/2) \end{bmatrix}.$$

(Review Example 7A.6.)

We can now make a schematic diagram to model the Hilbert space description of measuring for components of electron spin in two different directions, as we did in Figure 3A.7 in Chapter 3. Refer now to Figure 8C.3.

If we pick an orthogonal pair of one-dimensional subspaces to represent z-spin, then we can determine where to place an orthogonal pair of subspaces to represent w-spin for a direction w determined by an angle θ in Figure 8C.2. For simplicity we are considering $\varphi = 0$. The subspace corresponding to w-spin-up should be determined by the vector

$$\chi_w^{\mathrm{up}} = \begin{bmatrix} 1 \\ 0 \end{bmatrix}_w = \begin{bmatrix} \cos(\theta/2) \\ \sin(\theta/2) \end{bmatrix}$$

and the subspace corresponding to w-spin-down should be determined by the vector

$$\chi_w^{\mathrm{dn}} = \begin{bmatrix} 0 \\ 1 \end{bmatrix}_w = \begin{bmatrix} -\sin(\theta/2) \\ \cos(\theta/2) \end{bmatrix}.$$

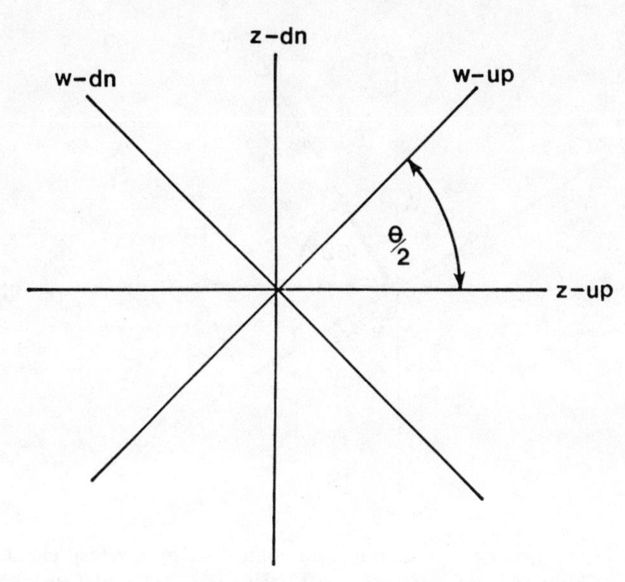

Figure 8C.3. Two orthogonal pairs of subspaces representing experiments to measure electron spin in two different directions.

(This explains the 45-degree angle in Figure 3A.7, because there $\theta = 90$ degrees.)

Let us apply this model to an example which we shall use later. Suppose $\theta = 120$ degrees. Then if an electron is in state $\begin{bmatrix} 1 \\ 0 \end{bmatrix}_z$, z-spin-up with certainty, what is the probability that a measurement for w-spin will yield outcome w-spin-up? In other words, what is the square of the component of $\begin{bmatrix} 1 \\ 0 \end{bmatrix}_z$ in the $\begin{bmatrix} 1 \\ 0 \end{bmatrix}_w$ direction? See Figure 8C.4.

We write

$$\begin{bmatrix} 1 \\ 0 \end{bmatrix}_z = \alpha \begin{bmatrix} 1 \\ 0 \end{bmatrix}_w + \beta \begin{bmatrix} 0 \\ 1 \end{bmatrix}_w = \alpha \begin{bmatrix} \cos(\theta/2) \\ \sin(\theta/2) \end{bmatrix} + \beta \begin{bmatrix} -\sin(\theta/2) \\ \cos(\theta/2) \end{bmatrix},$$

and solve for α and β to find that $\beta = \sin(\theta/2)$ and $\alpha = \cos(\theta/2)$. Thus, the probability that a measurement for w-spin will yield outcome w-spin-up is $\alpha^2 = \cos^2(60°) = \frac{1}{4}$, and the probability that the measurement will yield outcome w-spin-down is $\beta^2 = \sin^2(60°) = \frac{3}{4}$. Notice that the state $\begin{bmatrix} 1 \\ 0 \end{bmatrix}_z$ is one of certainty with respect to z-spin while at the same time it is a state of uncertainty with respect to w-spin. As we discussed in Chapter 3, it is the act of measurement that *puts* the electron in a dispersion-free state with respect to the observable measured. Thus, a measurement for w-spin would

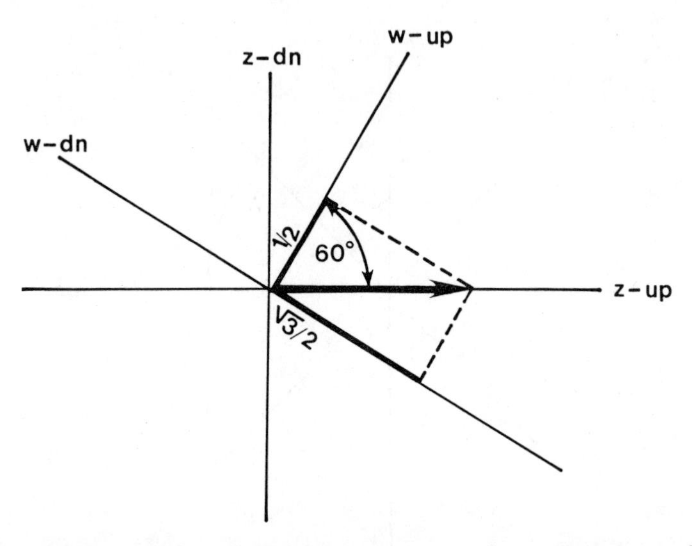

Figure 8C.4. The probability of obtaining outcome *w*-up for a measurement of *w*-spin while electron is in state *z*-up is determined by the projection of the state vector onto the *w*-up subspace.

put the electron in a dispersion-free *w*-spin state, which would have *z*-spin components that were not dispersion-free.

We turn next to a two-particle physical system. An example of such a system is a pair of electrons in a helium atom. The complete quantum mechanical description of such a system is very complicated, but if we restrict our attention only to the spin variable, we can investigate an idea that is simple, yet profound. The mathematics used to join the two separate particles into a single system is the "direct product" or "outer product." We present here only the barest essentials of the mathematics for a two-electron system.

We define the *outer product of two vectors* $u = \begin{bmatrix} x \\ y \end{bmatrix}$, $v = \begin{bmatrix} m \\ n \end{bmatrix}$ by

$$\underline{u \otimes v} = \begin{bmatrix} x \\ y \end{bmatrix} \otimes \begin{bmatrix} m \\ n \end{bmatrix} = \begin{bmatrix} xm \\ xn \\ ym \\ yn \end{bmatrix}.$$

Thus, the product is a vector in four-dimensional space.

We define the *outer product of two matrices* $A = \begin{bmatrix} a & b \\ c & d \end{bmatrix}$ and $B = \begin{bmatrix} p & q \\ r & s \end{bmatrix}$ by

$$\underline{A \otimes B} = \begin{bmatrix} aB & bB \\ cB & dB \end{bmatrix} = \begin{bmatrix} ap & aq & bp & bq \\ ar & as & br & bs \\ cp & cq & dp & dq \\ cr & cs & dr & ds \end{bmatrix}.$$

It is easy to check that $(A \otimes B)(u \otimes v) = Au \otimes Bv$, a desirable property for a two-particle system.

If we identify our two particles in the system as particle 1 and particle 2, then for direction z in space, the operators representing z-spin for particles 1 and 2, respectively, are (refer to equation 8C.2)

$$\underline{S_{1z}} = S_z \otimes I = \frac{1}{2}\begin{bmatrix} 1 & 0 \\ 0 & -1 \end{bmatrix} \otimes \begin{bmatrix} 1 & 0 \\ 0 & 1 \end{bmatrix} = \frac{1}{2}\begin{bmatrix} 1 & 0 & 0 & 0 \\ 0 & 1 & 0 & 0 \\ 0 & 0 & -1 & 0 \\ 0 & 0 & 0 & -1 \end{bmatrix},$$

$$\underline{S_{2z}} = I \otimes S_z = \begin{bmatrix} 1 & 0 \\ 0 & 1 \end{bmatrix} \otimes \frac{1}{2}\begin{bmatrix} 1 & 0 \\ 0 & -1 \end{bmatrix} = \frac{1}{2}\begin{bmatrix} 1 & 0 & 0 & 0 \\ 0 & -1 & 0 & 0 \\ 0 & 0 & 1 & 0 \\ 0 & 0 & 0 & -1 \end{bmatrix}. \tag{8C.3}$$

The *total z-spin* for the two-particle system is defined by operator

$$\underline{S_z} = S_{1z} + S_{2z} = \begin{bmatrix} 1 & 0 & 0 & 0 \\ 0 & 0 & 0 & 0 \\ 0 & 0 & 0 & 0 \\ 0 & 0 & 0 & -1 \end{bmatrix}.$$

The *total z-spin state functions* for the system are eigenvectors of S_z, and their corresponding eigenvalues are the magnitude of the total z-spin for the system. Notice that

$$\begin{bmatrix} 0 \\ 1 \\ 0 \\ 0 \end{bmatrix} = \begin{bmatrix} 1 \\ 0 \end{bmatrix} \otimes \begin{bmatrix} 0 \\ 1 \end{bmatrix} \quad \text{and} \quad \begin{bmatrix} 0 \\ 0 \\ 1 \\ 0 \end{bmatrix} = \begin{bmatrix} 0 \\ 1 \end{bmatrix} \otimes \begin{bmatrix} 1 \\ 0 \end{bmatrix}$$

are eigenvectors of S_z with eigenvalue zero. They represent two states in which the two electrons have opposite z-spins.

We can make a schematic sketch to consider these states in a four-dimensional Hilbert space similar to the sketch in Figure 3A.7. In Figure 8C.5 we have shown two dimensions in a four-dimensional Hilbert space. We have also sketched a state

$$\psi = \frac{1}{\sqrt{2}}\left(\begin{bmatrix} 1 \\ 0 \end{bmatrix} \otimes \begin{bmatrix} 0 \\ 1 \end{bmatrix} - \begin{bmatrix} 0 \\ 1 \end{bmatrix} \otimes \begin{bmatrix} 1 \\ 0 \end{bmatrix}\right) = \frac{1}{\sqrt{2}}\begin{bmatrix} 0 \\ 1 \\ -1 \\ 0 \end{bmatrix}, \tag{8C.4}$$

called the *singlet state* of the two-electron system, which will be important in the discussion to follow. We now have the tools to set up an EPR apparatus.

In 1935 in *Physical Review* (Vol. 47, p. 777) Albert Einstein, Boris Podolsky, and Nathan Rosen published a paper entitled "Can Quantum Mechanical

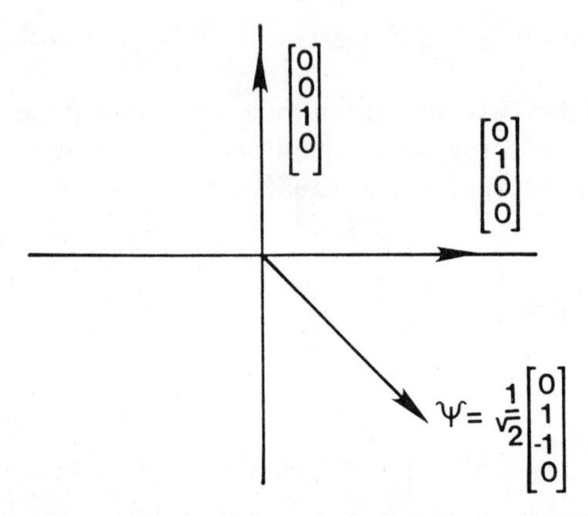

Figure 8C.5. A representation of a spin state in four dimensions for a two-particle system.

Description of Physical Reality Be Considered Complete?" The paper is commonly referred to as "the EPR paper" after the names of its authors. Their purpose was to prove that quantum mechanics could not be considered a complete description of the universe. They presented a "thought experiment," an experiment on a two-particle system that could not be performed at the time but whose outcomes were predicted by quantum mechanics. In the past 50 years hundreds of books and papers have been written about the "EPR dilemma." Below we give a fascinatingly simple and elegant presentation of the EPR thought experiment based on the work of N. David Mermin.

We consider a system of two electrons in a single orbit in an atom. We suppose the system is in the singlet state ψ given by equation (8C.4). We then arrange to remove the electrons from their orbit and send them flying off in opposite directions along straight lines, as pictured in Figure 8C.6. This is

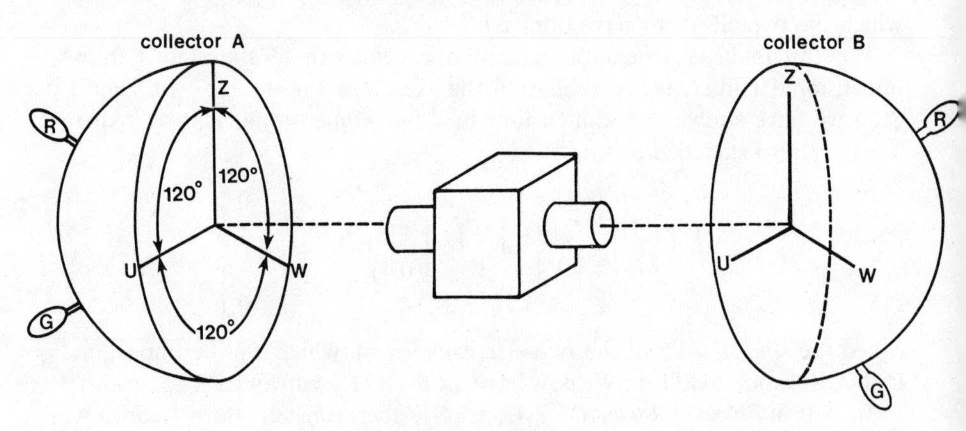

Figure 8C.6. An EPR apparatus.

the "thought" part of the experiment. It was not practically possible to do this in 1935, although today modern laboratory experiments are beginning to verify the quantum mechanics predictions cited in the EPR paper.

Suppose now that a collector containing a Stern–Gerlach apparatus is set up in a region A in the path of particle 1 (moving to the left in Figure 8C.6) and another in a region B in the path of particle 2. Suppose also that the collectors are very far apart, relative to the size of the electrons. We arrange it so that each collector is capable of measuring for the spin of the electron in one of three spatial directions, z, u, w, each direction making a 120-degree angle with the other two. The direction z is the same spatial direction for both collectors, and the same is true for directions u and w. Finally, we rig up two light bulbs, a green and a red, at each collector. We wire collector A so that for each of the three directions, if the collector measures an electron to have spin *up* in that direction, the *red* bulb will flash. If spin *down*, the *green* bulb will flash. We wire collector B just the opposite—if for any of the three directions the collector measures an electron to have spin *up*, the *green* bulb will flash; for spin *down*, the *red* bulb.

We now start firing pairs of electrons, all in the singlet state. During these firings we randomly change each of the collectors independently to measure for spin in the z, u, and w directions. For each firing we record the color of the flash at A and the color at B. What does quantum mechanics predict about the results of our experiment?

Suppose first that we set both collectors to measure for spin in the same direction, say z. We fire an electron pair, and we call the electron that arrives at collector A the "left" electron and the other the "right" electron. Suppose we get a green flash at A. Then we know that the left electron is in state $\begin{bmatrix} 0 \\ 1 \end{bmatrix}_z$, so that the measurement has put the pair in the state

$$\begin{bmatrix} 0 \\ 0 \\ 1 \\ 0 \end{bmatrix}_z = \begin{bmatrix} 0 \\ 1 \end{bmatrix}_z \otimes \begin{bmatrix} 1 \\ 0 \end{bmatrix}_z .$$

(Recall that the component of the singlet state with respect to this dispersion-free state was $1/\sqrt{2}$, indicating that we had a 50 percent probability of obtaining a green flash instead of a red flash.) What color will flash at collector B? Since the z-spin for the right particle is up, collector B will also flash a green bulb. On the other hand, if we had seen a red flash at A, then the measurement must have put the pair in state $\begin{bmatrix} 1 \\ 0 \end{bmatrix}_z \otimes \begin{bmatrix} 0 \\ 1 \end{bmatrix}_z$, so that we would also see a red flash at B. In summary:

> *if we set both collectors to measure spin in direction z, then the collectors will flash the same color for every electron pair fired.*

At this stage we see already one of the issues so troublesome to Einstein, Podolsky, and Rosen. The quantum mechanical description of this experiment holds that a measurement at collector A *puts* the system in a certain state and thus affects what an observer will see at B, which is at great distance from A. A dilemma, therefore, arises if on one hand the z-spins of the electrons are considered "local properties," while on the other hand measurements for these properties send correlation signals over great distances faster than the speed of light! The EPR authors considered this "spooky action at a distance" a serious flaw in quantum mechanics. We are about to see, however, that the EPR apparatus produces an even more dramatic dilemma.

Suppose now that we set collector A to measure for spin in direction z and collector B to measure for spin in direction w making an angle $\theta = 120$ degrees with the z direction. (Suppose $\varphi = 0$ degrees.) If we fire an electron pair and collector A flashes green, then we have put the system in state $\begin{bmatrix} 0 \\ 1 \end{bmatrix}_z \otimes \begin{bmatrix} 1 \\ 0 \end{bmatrix}_z$, as before. What light will flash at B? Since the right particle is in state $\begin{bmatrix} 1 \\ 0 \end{bmatrix}_z$, we see from the calculation following Figure 8C.4 that the probability of obtaining a green flash (w-spin up) is $\frac{1}{4}$ and the probability of obtaining a red flash (w-spin down) is $\frac{3}{4}$. Of course, the same probabilities for green and red flashes at B obtain if we measure for u-spin at B, since the u direction also is at an angle of 120 degrees with the z axis. (The nonzero value for φ for direction u need not complicate matters, because our orientation in space is arbitrary and the argument for u is exactly symmetric with our argument for w.)

We are now ready to summarize the quantum mechanics prediction of the results of running 900 electron pairs through the EPR apparatus, 100 for each of the 9 possible pairs of collector settings. (The number 900 is not important; we could use any number of electron pairs.) For each pair we record whether the bulbs flashed the same colors or different colors. The results are tabulated below:

Collector settings ⟶	zz	zu	zw	uz	uu	uw	wz	wu	ww
Same color →	100	25	25	25	100	25	25	25	100
Different →	0	75	75	75	0	75	75	75	0

The table shows two facts:

Observation 1: When the collectors are set in the same direction, the two bulbs always flash the same color, a consequence of the fact that the singlet state is antisymmetric.

Observation 2: Over all nine possible settings for the two collectors, the two bulbs flash the same color for exactly half (450) the total number of electron pairs fired, a consequence of the angle $\theta = 120$ degrees between directions.

We shall see a remarkable dilemma arise when we try to account for these observations by supposing that electron spin is an element of "reality" that exists *before* it is measured.

If we suppose that the two electrons heading for the two collectors already "have" spin components in z, w, and u directions, then each electron can be considered to be carrying a code that will tell how the collectors should flash for each of the three directions. For example, the left electron might carry the code RRG, indicating that in directions z, w, and u, respectively, the spin components are z-up (red), w-up (red), and u-down (green). If the electrons carry such codes, then observation 1 above forces us to the conclusion that *both electrons in a pair must always carry the same code.* For if the left carried RRG and the right RGG, for example, then the collectors would flash different colors when both are set in direction w, contradicting observation 1.

Now consider the following table, in which we indicate, over all possible collector settings and all possible codes the electrons may carry, whether or not the bulbs will flash the same color (S) or different colors (D).

Collector settings → / Codes ↓ z-u-w	zz	zu	zw	uz	uu	uw	wz	wu	ww
R R R	S	S	S	S	S	S	S	S	S
R R G	S	S	D	S	S	D	D	D	S
R G R	S	D	S	D	S	D	S	D	S
R G G	S	D	D	D	S	S	D	S	S
G R R	S	D	D	D	S	S	D	S	S
G R G	S	D	S	D	S	D	S	D	S
G G R	S	S	D	S	S	D	D	D	S
G G G	S	S	S	S	S	S	S	S	S

The table reveals a startling fact. Over all possible settings of the collectors and over all possible codes, the bulbs must flash the *same color at least five-ninths of the time*—that is, for at least 500 out of the 900 pairs fired. This is because no matter which code an electron pair carries, at least five of the nine settings of the collectors will result in flashes of the same color. But this is a contradiction to observation 2 above.

Are we forced to conclude that codes do not exist—that electron spin is not a reality until it is measured? Almost, but not quite. There is one escape hatch. It could be the case that codes exist but that a measurement at A changes the code. But then it must change the code for both electrons simultaneously, because the measurements at A and B could occur simultaneously. Given the possibility of setting up collectors at a great distance from each other, this requires a passing of information from A to B at speeds much greater than the speed of light, contradicting relativity theory.

This then is the EPR paradox:

> Either electron spin before it is measured is an element of "physical reality" *necessarily absent* from the quantum mechanical description of the universe, or else measurements at one location can influence events at other distant locations with a speed greater than the speed of light.

The EPR paradox was not resolved in Einstein's lifetime, and so this most revered thinker died at odds with the majority of physicists of his time, clinging to the view that physical "reality" exists outside the domain of measurements by humans. Of course, Einstein's view is the one defendable by the great traditions of Western thought, including Platonism, Western religions, and classical physics among others. It is understandable that many people would oppose throwing out all of these merely because a peculiar model for describing subatomic phenomena is enjoying a high predictive value at the moment—even if its predictive value seems to increase with many new experiments introduced every year. This is one reason behind much of the research in quantum logic. Perhaps the day will come when a coherent theory of quantum physics will emerge incorporating all the essential *logical* features of noncompatibility of observables and all the computational features of orthodox quantum mechanics, but unencumbered by the physically mysterious axioms of Hilbert space that lead to the EPR dilemma and other quandaries. It is little wonder that quantum mechanics is the object of such intense scrutiny and debate in the worlds of science, mathematics, and philosophy.

P. viii

Coaching Manual

Chapter 1A

1A.3. Project

$$\operatorname{Exp}(E, \omega) = \sum_{x=1}^{20} x\omega(x) = \frac{1}{300}(0 + 1{\cdot}8 + 2{\cdot}12 + 3{\cdot}11 \text{ etc.}) = \frac{3491}{300}$$
$$= 11.64$$

1A.4. Project

Suppose $k \in \mathbb{N}$ with $0 \leqq k \leqq 40$. Then

$$\mu(J_k) = \begin{cases} 0, & \text{if } k \text{ is odd,} \\ \omega(k/2) & \text{if } k \text{ is even} \end{cases}$$

because, for example, no outcomes of E are in the intervals $J_1 = (0, 1)$ or $J_5 = (4, 5)$, while the only outcome in $J_6 = \{3\}$ is $x = 3$. Also note that $\mu(J_{20}) = \omega(20)$, because 20 is the only outcome in $J_{20} = \{20\}$. The required verification now follows easily.

1A.6. Lemmas

A. Suppose $\mu(A) < \infty$. Then $\mu(\varnothing) = \mu(\varnothing \cup \varnothing) = \mu(\varnothing) + \mu(\varnothing)$. So $\mu(\varnothing) = 0$.

B. (i) We can write B as a disjoint union: $B = A \cup B \backslash A$. Then $\mu(B) = \mu(A) + \mu(B \backslash A)$, from which the inequality follows.

(ii) Suppose $\langle\langle A_n \rangle\rangle$ is an increasing sequence in $\mathbb{A}$. If $\mu(A_n) = \infty$ for some n, then clearly the conclusion of the lemma is true. So let us suppose that $\mu(A_n) < \infty$ for all n. For each $n \in \mathbb{N}$, with $n > 1$, let

$B_n = A_n \backslash A_{n-1}$. Then $\langle\langle B_n \rangle\rangle$ is a pairwise disjoint sequence, so for each n, $\mu(A_n) = \sum_{k=1}^{n} \mu(B_k)$. We then have that

$$\mu\left(\bigcup_n A_n\right) = \mu\left(\bigcup_n B_n\right) = \sum_{n=1}^{\infty} \mu(B_n) = \sup_n \sum_{k=1}^{n} \mu(B_k) = \sup_n \mu(A_n).$$

(iii) Without loss in generality, we can assume $\mu(A_1) < \infty$. If $B \subseteq A_1$ and $B \in \mathbb{A}$, then $(A_1 \backslash B) \cap B = \varnothing$, so that $\mu(A_1 \backslash B) + \mu(B) = \mu(A_1)$, from which it follows that

$$\mu(A_1 \backslash B) = \mu(A_1) - \mu(B). \tag{$*$}$$

Since $\bigcap_n A_n \subseteq A_1$, we have $\mu(A_1) - \mu(\bigcap_n A_n) = \mu(A_1 \backslash \bigcap_n A_n) = \mu(\bigcup_n A_1 \backslash A_n) = \sup_n \mu(A_1 \backslash A_n)$ from (ii). Then, using ($*$), we have $\mu(A_1) - \mu(\bigcap_n A_n) = \sup_n (\mu(A_1) - \mu(A_n)) = \mu(A_1) - \inf_n \mu(A_n)$, from which the conclusion of the lemma follows.

C. The only subtlety is showing that there exists a set of finite measure. But if $\mu_1(A) < \infty$ and $\mu_1(B) < \infty$, then $\mu_1(A \cap B) < \infty$ by 1A.6B(i). Since there exists A with $\mu_1(A) < \infty$ and there exists B with $\mu_2(B) < \infty$, we then have that $(\mu_1 + \mu_2)(A \cap B) < \infty$.

1A.8. Lemmas

A. The proof consists of 10 equalities, each proving that the set on the left of the equality is a Borel set by writing it as a countable union or difference of Borel sets. The fact that the sets on the right of an equality are Borel sets often depends on a previous equality in the order we present here.

$$(-\infty, a] = \bigcup_{k \in \mathbb{N}} (a-k, a];$$

$$(b, \infty) = \mathfrak{R} \backslash (-\infty, b]; \qquad (-\infty, a) = \bigcup_{k \in \mathbb{N}} \left(-\infty, a - \frac{1}{k}\right];$$

$$[b, \infty) = \mathfrak{R} \backslash (-\infty, b); \qquad (a, b] = (-\infty, b] \backslash (-\infty, a];$$

$$(a, b) = (-\infty, b) \backslash (-\infty, a];$$

$$[a, b) = (-\infty, b) \backslash (-\infty, a); \qquad [a, b] = (-\infty, b] \backslash (-\infty, a);$$

$$\{a\} = [a, a]; \qquad \mathfrak{R} = (-\infty, a] \cup (a, \infty); \qquad \varnothing = \mathfrak{R} \backslash \mathfrak{R}.$$

B. Since we proved in A that $\mathfrak{R} \in \mathbb{B}$, it is then obvious that $\mathbb{B}$ satisfies (i)–(iii) in Definition 1A.5 of a σ-algebra.

1A.10. Examples

A. (i) $\mu(x, \infty) = \mu(\bigcup_{k=0}^{\infty} (x+k, x+k+1]) = \sum_{k=0}^{\infty} \mu(x+k, x+k+1] = \sum_{k=0}^{\infty} 1 = \infty$. The other equality is proved analogously.

(ii) $\mu\{x\} = \mu[x, x] = x - x = 0$.

(iii) If C is countable, C consists of a countable union of singletons. The union is pairwise disjoint, and by (ii) each singleton has Lebesgue measure zero.

B. It is obvious that the properties in Definition 1A.5 (i′) and (ii′) are satisfied. The fact that (iii′) is also satisfied follows from the fact that if $\mathbb{S}$ is a pairwise disjoint, countable collection of Borel sets, then x belongs to at most one $S \in \mathbb{S}$.

1A.16. Examples

A. The set $\{x \mid f(x) \neq 0\}$ has Lebesgue measure zero because it is the countable set of rational numbers between 0 and 1.

B. $f(x) = \begin{cases} x & \text{if } x \text{ is irrational,} \\ 0 & \text{if } x \text{ is rational.} \end{cases}$

1A.18. Lemma

It is easy to establish that f^+ and f^- are nonnegative functions and that $f = f^+ - f^-$.

To see that f^+ is μ-measurable, suppose $B \in \mathbb{B}$. We must show that $(f^+)^{\leftarrow}[B] \in \mathbb{A}$.

In the case $0 \in B$, you should be able to show that $(f^+)^{\leftarrow}[B] = f^{\leftarrow}[B \cap [0, \infty)] \cup f^{\leftarrow}[(-\infty, 0]]$, which is the union of two sets belonging to $\mathbb{A}$, because f is μ-measurable.

In the case $0 \notin B$, then $(f^+)^{\leftarrow}[B] = f^{\leftarrow}[B \cap [0, \infty)]$, which belongs to $\mathbb{A}$, because f is μ-measurable.

The proof that f^- is μ-measurable is similar.

Chapter 1B

1B.6. Theorem

B. Our proof will rest on the following claim:

Claim. If μ is a measure on σ-algebra $\mathbb{A}$ in $\mathfrak{R}$, $S \in \mathbb{A}$, and p and q are nonnegative functions that are μ-integrable over S, and if $p(x) \leqq q(x)$ for all x in S, then $\int_S p \, d\mu \leqq \int_S q \, d\mu$.

Before we turn to the proof of the claim, let us see how it will solve our problem. Our task is to show that

$$\int_S f^+ \, d\mu - \int_S f^- \, d\mu \leqq \int_S g^+ \, d\mu - \int_S g^- \, d\mu. \tag{$*$}$$

Since $f^+(x) \leqq g^+(x)$ and $g^-(x) \leqq f^-(x)$ for all x in S, our claim implies that

$$\int_S f^+ \, d\mu \leqq \int_S g^+ \, d\mu \quad \text{and} \quad \int_S g^- \, d\mu \leqq \int_S f^- \, d\mu,$$

and all integrals are finite. From this, inequality $(*)$ is easily established.

It remains to prove the claim. Let

$$P = \left\{ \int_S h \, d\mu \,\middle|\, h \text{ is a nonnegative, } \mu\text{-measurable, simple function with } h(x) \leqq p(x) \text{ for all } x \text{ in } S. \right\}$$

Let

$$Q = \left\{ \int_S j \, d\mu \,\middle|\, j \text{ is a nonnegative, } \mu\text{-measurable, simple function with } j(x) \leqq q(x) \text{ for all } x \text{ in } S. \right\}$$

Show that every number in P is also in Q. This will establish that $\sup(P) \leqq \sup(Q)$. This will complete the proof of Theorem 1B.6B.

1B.8. Theorems

A. Let μ be a measure on $\mathbb{B}$, and let S be a μ-measurable set. Suppose f is a function that is μ-integrable over S and $f(x) = 0$ μ-ae on S.

First, we show that $\int_S f^+ \, d\mu = 0$, as follows. Let $T = (x \in S \mid f^+(x) \neq 0)$. Then $T \subseteq (x \in S \mid f(x) \neq 0)$. Suppose h is a simple μ-measurable function with $0 \leqq h(x) \leqq f^+(x)$ for all x in S, and suppose image $(h) = (a_1, \ldots, a_n)$. Show that for each k, if $a_k \neq 0$, then $(h^{\leftarrow}[a_k] \cap S) \subseteq T$. Then show that this implies that $\int_S h \, d\mu = 0$. From the fact that h was arbitrary, we have that $\int_S f^+ \, d\mu = 0$. A similar proof shows $\int_S f^- \, d\mu = 0$.

B. Let f be a nonnegative μ-measurable function and S a μ-measurable set with $\int_S f \, d\mu = 0$. For each natural number n define $A_n = \{x \in S \mid 1/n < f(x)\}$. Let $T = \bigcup_{n=1}^{\infty} A_n$. Our problem will be solved if we can show that $\mu(T) = 0$. By Lemma 1A.6B(ii) it suffices to show that $\mu(A_n) = 0$ for every n. For every natural number n, consider the simple function

$$h(x) = \begin{cases} 0 & \text{if } x \in S \setminus A_n, \\ 1/n & \text{if } x \in A_n. \end{cases}$$

Show that h is a μ-measurable simple function and that $h(x) \leqq f(x)$ for all x in S. Hence, by Theorem 1B.6B, we have that $\int_S h\,d\mu \leqq \int_S f\,d\mu$. Then show that the integral on the left is $\mu(A_n)/n$, and the integral on the right is zero.

1B.9. Examples

A. The function f_0 is a simple function.
B. Find the measure of the set $T = \{x \mid x \in I \text{ and } f_1(x) \neq 0\}$.
C. Find the measure of the set $T = \{x \mid x \in S \text{ and } x \neq 2\}$. The value of the integral is 2π.

1B.10. Lemma

For $k = 0, \ldots, 40$ define

$$J_k = \begin{cases} \{k/2\} & \text{if } k \text{ is even,} \\ \left(\dfrac{k-1}{2}, \dfrac{k+1}{2}\right) & \text{if } k \text{ is odd.} \end{cases}$$

Define simple function f by

$$f(x) = \begin{cases} k/2 & \text{if } k \text{ is even and } x \in J_k, \\ 0 & \text{otherwise.} \end{cases}$$

Then $\mu(20, \infty) = \mu(-\infty, 0) = 0$, and for each odd k, $\mu(J_k) = 0$. This establishes that $f = I_{\mathfrak{R}}$ μ-ae, hence that $\int_{\mathfrak{R}} f\,d\mu = \int_{\mathfrak{R}} I_{\mathfrak{R}}\,d\mu$. The proof is then completed by observing that the integral on the left of this last equality is the sum on the right of the equality in equation (1A.2).

Chapter 2

2.2. Lemma

$$\langle x, \lambda y \rangle = \langle \lambda y, x \rangle^* = (\lambda \langle y, x \rangle)^* = \lambda^* \langle y, x \rangle^* = \lambda^* \langle x, y \rangle.$$

2.3. Examples and Projects

A. It is not difficult to show from properties of complex arithmetic that $\langle , \rangle$ satisfies properties 2.1(ii)(a)–(d). To establish property (d), for example, observe that for every complex number x_1, $x_1 x_1^* = |x_1|^2 \geqq 0$, and equality holds only if $x_1 = 0$.
B. To see that the series $\sum_{k=1}^{\infty} x_k y_k^*$ converges, note that for every $n \in \mathbb{N}$, $0 \leqq (|x_n| - |y_n|)^2 = |x_n|^2 - 2|x_n|\,|y_n| + |y_n|^2$, hence $2|x_n y_n^*| = 2|x_n|\,|y_n^*| =$

$2|x_n|\,|y_n| \leqq |x_n|^2 + |y_n|^2$. Thus, $\sum_{k=1}^{\infty} |x_k y_k^*| \leqq \frac{1}{2}(\sum_{k=1}^{\infty} |x_n|^2 + \sum_{k=1}^{\infty} |y_n|^2)$, and both series on the right converge because $x, y \in V$. Since absolute convergence implies convergence in $\mathfrak{C}$, the proof is complete.

That $\langle \, , \rangle$ satisfies the properties of an inner product follows from 2.3A above and standard theorems about convergent series in $\mathfrak{C}$.

C. To show that pointwise addition and scalar multiplication can be transferred from V to $\hat{V}$ we first prove a lemma.

Lemma. *If $\psi, \psi', \gamma, \gamma' \in V$, and $\psi = \psi'$ μ-ae on S, and $\gamma = \gamma'$ μ-ae on S, then $\psi + \gamma = \psi' + \gamma'$ μ-ae on S.*

PROOF. Let $A = \{t \in S \mid \psi(t) \neq \psi'(t)\}$, and $B = \{t \in S \mid \gamma(t) \neq \gamma'(t)\}$, and $C = \{t \in S \mid \psi(t) + \gamma(t) \neq \psi'(t) + \gamma'(t)\}$. Then $C \subseteq A \cup B$, so $\mu(C) \leqq \mu(A \cup B) \leqq \mu(A) + \mu(B)$. But the μ-measures of both A and B are zero by hypothesis, so the lemma is proved. □

Now let the equivalence class in $\hat{V}$ generated by $\psi \in V$ be denoted by $[\psi]$. We define pointwise addition on $\hat{V}$ as follows: for $[\psi]$, $[\gamma] \in \hat{V}$, define $[\psi] + [\gamma] = [\psi + \gamma]$.

To show that these operations are well defined we must show that if $[\psi] = [\psi']$ and $[\gamma] = [\gamma']$, then $[\psi + \gamma] = [\psi' + \gamma']$. But this is exactly what our lemma says, because $[\psi] = [\psi']$ if and only if $\psi = \psi'$ μ-ae on S.

A similar argument is used to define scalar multiplication on $\hat{V}$ by $\lambda[\psi] = [\lambda\psi]$ for $\lambda \in \mathfrak{C}$ and $[\psi] \in \hat{V}$.

The inner product on $\hat{V}$ is defined for all $[\psi_1], [\psi_2] \in \hat{V}$ by

$$\langle [\psi_1], [\psi_2] \rangle = \int_S \psi_1 \psi_2^* \, d\mu.$$

Again, we must show that this operation is well defined. If $[\psi_1] = [\psi'_1]$ and $[\psi_2] = [\psi'_2]$, then $\psi_2^* = (\psi'_2)^*$ μ-ae on S. From our lemma it then follows that $\psi_1 \psi_2^* - \psi'_1(\psi'_2)^* = 0$ μ-ae on S, and so $\int_S \psi_1 \psi_2^* \, d\mu = \int_S \psi'_1 (\psi'_2)^* \, d\mu$ by Theorem 1B.8A.

It is now a routine exercise in complex arithmetic to verify that the inner product that we have defined on $\hat{V}$ satisfies the properties in Definition 2.1(ii).

2.5. Project

This is a routine computation in $\mathfrak{C}$ using the fact that for $t = a + ib \in \mathfrak{C}$, $tt^* = a^2 + b^2 = |t|^2$.

2.6. Theorem

(i) Since $\langle x, x \rangle$ is a nonnegative real number, so is its square root. If the square root is zero, then so is $\langle x, x \rangle$, which implies that x is the zero vector by the properties of the inner product.

(ii) This follows easily from a routine computation using the properties of inner products.
(iii) This also follows easily from computations using the fact that $\langle x, -y\rangle = (-1)^* \langle x, y\rangle = -1\langle x, y\rangle$.

2.8. Lemma

$$\langle x, 0\rangle = \langle x, x - x\rangle = \langle x, x\rangle + \langle x, -x\rangle = \langle x, x\rangle - \langle x, x\rangle = 0.$$

2.11. Theorem (Bessel's Inequality)

Suppose $\{x_1, \ldots, x_p\}$ is an orthonormal set in V. Let $z = y - \sum_{k=1}^{p} \langle x_k, y\rangle x_k$. Then for $k = 1, \ldots, p$ we have that $z \perp \langle x_k, y\rangle x_k$. This follows from a straightforward computation using the fact that $\langle x_k, x_k\rangle = 1$ for all k and $\langle x_k, x_j\rangle = 0$ for all $j \neq k$. Then by the Pythagorean theorem we have $\|y\|^2 = \|z + \sum_{k=1}^{p} \langle x_k, y\rangle x_k\|^2 = \|z\|^2 + \sum_{k=1}^{p} \|\langle x_k, y\rangle x_k\|^2 = \|z\|^2 + \sum_{k=1}^{p} |\langle x_k, y\rangle|^2 \geqq \sum_{k=1}^{p} |\langle x_k, y\rangle|^2$. The last equality follows from the fact that for every k, $\|\langle x_k, y\rangle x_k\|^2 = |\langle x_k, y\rangle|^2 \|x_k\|^2$, and the norm of x_k equals one.

2.12. Corollaries to Bessel's Inequality

A. If $\langle\langle x_k \rangle\rangle$ is an orthonormal sequence in V, we conclude from Bessel's inequality that the nondecreasing sequence of partial sums for the nonnegative real series $\sum_{k=1}^{\infty} |\langle y, x_k\rangle|^2$ is bounded above by $\|y\|^2$.
B. Apply Bessel's inequality to the orthonormal set $\{x\}$.

2.13. Theorem (The Cauchy–Schwarz Inequality)

If $x = 0$, the inequality clearly holds. For $x \neq 0$, $\|(1/\|x\|)x\| = 1$. So by Corollary 2.12B,

$$\frac{1}{\|x\|} |\langle x, y\rangle 1 = \left| \left\langle \frac{x}{\|x\|}, y \right\rangle \right| \leqq \|y\|.$$

2.14. Theorems

A. $\|x + y\|^2 = |\langle x + y, x + y\rangle| \leqq \|x\|^2 + |\langle x, y\rangle| + |\langle y, x\rangle| + \|y\|^2 \leqq \|x\|^2 + 2\|x\| \|y\| + \|y\|^2 = (\|x\| + \|y\|)^2$. The third step follows from Theorem 2.13.
B. $\|x\| = \|(x - y) + y\| \leqq \|x - y\| + \|y\|$ from the triangle inequality.

Similarly, $\|y\| \leqq \|x-y\| + \|x\|$. So $\|x-y\| \geqq \|x\| - \|y\|$ and $\|x-y\| \geqq -(\|x\| - \|y\|)$, from which the conclusion of the theorem follows.

2.16. Theorem

Suppose that $\langle\langle x_k \rangle\rangle$ converges in norm to $x \in V$. Suppose $\varepsilon > 0$. Then there exists N_ε such that for all $k > N_\varepsilon$, $\|x_k - x\| < \varepsilon/2$. Then for all $k, j > N_\varepsilon$, $\|x_k - x_j\| = \|x_k - x + x - x_j\| \leqq \|x_k - x\| + \|x - x_j\| \leqq \varepsilon/2 + \varepsilon/2 = \varepsilon$.

2.19. Examples and Projects

A. To show that $\mathfrak{C}^n$ is complete, suppose $\langle\langle x_k \rangle\rangle$ is a Cauchy sequence in $\mathfrak{C}^n$ with $x_k = (x_k^1, \ldots, x_k^n)$ for every $k \in \mathbb{N}$. We will show that the coordinates of the x_k's form n Cauchy sequences of complex numbers, hence converge to n complex numbers, which then form the coordinates of the vector in $\mathfrak{C}^n$ to which $\langle\langle x_k \rangle\rangle$ converges. Here are the details.

For every $\varepsilon > 0$ there exists $N_\varepsilon \in \mathbb{N}$ such that for all $k, j > N_\varepsilon$ and for all $i = 1, \ldots, n$, $|x_k^i - x_j^i|^2 \leqq \sum_{p=1}^n |x_k^p - x_j^p|^2 = \|x_k - x_j\|^2 < \varepsilon$. This shows that for every $i = 1, \ldots, n$, the sequence $\langle\langle x_k^i \rangle\rangle$ $(k \in \mathbb{N})$ of complex numbers is Cauchy. Since the complex numbers form a complete field, there exists for each $i = 1, \ldots, n$ a complex number z^i such that $\lim_{k \to \infty} x_k^i = z^i$.

Now we shall show that $\langle\langle x_k \rangle\rangle$ converges in norm to $z = (z^1, \ldots, z^n)$. Given $\varepsilon > 0$, for each $i = 1, \ldots, n$ there exists $N_\varepsilon^i \in \mathbb{N}$ such that for all $k > N_\varepsilon^i$, $|x_k^i - z^i|^2 < \varepsilon/n$. Let $N_\varepsilon = \max\{N_\varepsilon^i \mid i = 1, \ldots, n\}$. Then for all $k > N_\varepsilon$, $|x_k^i - z^i|^2 < \varepsilon/n$ for all $i = 1, \ldots, n$. So for $k > N_\varepsilon$, we have $\|x_k - z\|^2 = \sum_{i=1}^n |x_k^i - z^i|^2 < \varepsilon$.

B. Suppose $\langle\langle x_k \rangle\rangle$ is a Cauchy sequence in l^2, and for all $k \in \mathbb{N}$, $x_k = (x_k^1, \ldots, x_k^i, \ldots)$. For every $i \in \mathbb{N}$, we can show in the same manner as in A above that $\langle\langle x_k^i \rangle\rangle$ is a Cauchy sequence in $\mathfrak{C}$ and so converges to a complex number z^i. Let $z = (z^1, \ldots, z^i, \ldots)$. We shall show that $z \in l^2$ and then that $\langle\langle x_k \rangle\rangle$ converges in norm to z.

Note first that for every $p \in \mathbb{N}$,

$$\lim_{j \to \infty} \sum_{i=1}^{p} |x_j^i - z^i|^2 = 0. \tag{$*$}$$

Now for every $\varepsilon > 0$, there exists $N_\varepsilon \in \mathbb{N}$ such that for all $k, j > N_\varepsilon$ and for all $p \in \mathbb{N}$, $\sum_{i=1}^p |x_k^i - x_j^i|^2 \leqq \sum_{i=1}^\infty |x_k^i - x_j^i|^2 = \|x_k - x_j\|^2 < \varepsilon$. So for fixed $k > N_\varepsilon$ and for all $p \in \mathbb{N}$,

$$\sum_{i=1}^{p} |x_k^i - z^i|^2 = \lim_{j \to \infty} \sum_{i=1}^{p} |x_k^i - z^i|^2 = \lim_{j \to \infty} \sum_{i=1}^{p} |x_k^i - x_j^i + x_j^i - z^i|^2 \tag{$**$}$$

$$\leqq \lim_{j \to \infty} \sum_{j=1}^{p} |x_k^i - x_j^i|^2 + \lim_{j \to \infty} \sum_{j=1}^{p} |x_j^i - z^i|^2 < \varepsilon + 0 = \varepsilon.$$

From this we conclude that for all $k > N_\varepsilon$, $\sum_{i=1}^{\infty} |x_k^i - z^i|^2 < \varepsilon$, so that $x_k - z \in l^2$. Hence, since $x_k \in l^2$ and $(l^2, +, \cdot)$ is a vector space, we have that $z = x_k - (x_k - z) \in l^2$.

Moreover, we conclude from (**) that

$$\lim_{k\to\infty} \|x_k - z\|^2 = \lim_{k\to\infty} \sum_{k=1}^{\infty} |x_k^i - z^i|^2 = 0,$$

which proves that $\langle\!\langle x_k \rangle\!\rangle$ converges in norm to z.

2.21. Theorems

A. From Theorem 2.14B, for every $k \in \mathbb{N}$, $|\,\|x\| - \|x_k\|\,| \leqq \|x - x_k\|$. Our hypothesis implies that $\lim_{k\to\infty} \|x - x_k\| = 0$.

B. (i). Using the Cauchy–Schwarz inequality, $|\langle x, y\rangle - \langle x_k, y_k\rangle| \leqq |\langle x, y\rangle - \langle x, y_k\rangle| + |\langle x, y_k\rangle - \langle x_k, y_k\rangle| = |\langle x, y - y_k\rangle| + |\langle x - x_k, y_k\rangle| \leqq \|x\|\,\|y - y_k\| + \|x - x_k\|\,\|y_k\|$. From part A of this theorem we know that $\lim_{k\to\infty} \|y_k\| = \|y\|$, so that the limit over k of both terms in this last sum is zero.

The proofs of (ii) and (iii) are routine results about sequences of complex numbers.

C. Let $x = \sum_{k=1}^{\infty} x_k$, and for every $n \in \mathbb{N}$, let $z_n = \sum_{k=1}^{n} x_k$ so that $\langle\!\langle z_n \rangle\!\rangle$ converges in norm to x. Then by the result in B above,

$$\sum_{k=1}^{\infty} \langle x_k, y\rangle = \lim_{n\to\infty} \sum_{k=1}^{n} \langle x_k, y\rangle = \lim_{n\to\infty} \left\langle \sum_{k=1}^{n} x_k, y \right\rangle = \lim_{n\to\infty} \langle z_n, y\rangle = \langle x, y\rangle.$$

D. Suppose $S = \{x_1, \ldots\}$ is an orthonormal set in H, and $x = \sum_{k=1}^{\infty} \lambda_k x_k$.

(i) For each $j \in \mathbb{N}$, $\lambda_j = \lambda_j \langle x_j, x_j\rangle = \sum_{k=1}^{\infty} \lambda_k \langle x_k, x_j\rangle = \langle \sum_{k=1}^{\infty} \lambda_k x_k, x_j\rangle = \langle x, x_j\rangle$. The first two equalities depend on the orthonormality, the third on part C of this theorem.

(ii) $\|x\|^2 = \langle x, x\rangle = \langle \sum_{k=1}^{\infty} \lambda_k x_k, \sum_{k=1}^{\infty} \lambda_k x_k\rangle = \sum_{k=1}^{\infty} \lambda_k \lambda_k^* \langle x_k, x_k\rangle = \sum_{k=1}^{\infty} |\lambda_k|^2$.

E. By the hypothesis, $\langle\!\langle \sum_{k=1}^{n} \|x_k\| \rangle\!\rangle$ $(n \in \mathbb{N})$ is a Cauchy sequence in $\mathbb{C}$. So for every $\varepsilon > 0$, there exists $N_\varepsilon \in \mathbb{N}$ such that for all $n, m > N_\varepsilon$, $\|\sum_{k=1}^{n} x_k - \sum_{k=1}^{m} x_k\| = \|\sum_{k=n}^{m} x_k\| \leqq \sum_{k=n}^{m} \|x_k\| < \varepsilon$, the first inequality following from Theorem 2.14A. Therefore, $\langle\!\langle \sum_{k=1}^{n} x_k \rangle\!\rangle$ $(n \in \mathbb{N})$ is a Cauchy sequence in H and hence converges. If S is pairwise orthogonal, then we obtain the third equality in the following string:

$$\|x\|^2 = \langle x, x\rangle = \left\langle \sum_{k=1}^{\infty} x_k, \sum_{k=1}^{\infty} x_k \right\rangle = \sum_{k=1}^{\infty} \langle x_k, x_k\rangle = \sum_{k=1}^{\infty} \|x_k\|^2.$$

2.23. Theorem

Suppose $S=\{x_1,\dots\}$ and $\sum_{k=1}^{\infty}\lambda_k x_k=0$. From Theorem 2.21D we see that for every $j\in\mathbb{N}, \lambda_j=\langle 0,x_j\rangle$, which equals zero by Lemma 2.8. Clearly, this same proof works if S is finite.

2.24 Theorem

We shall show that each part implies the next part and that the last implies the first.

(i)$\Rightarrow$(ii). Let B be a basis for H. Suppose $x\perp b$ for all $b\in B$. If $x\neq 0$, then $x/\|x\|$ is a unit vector not orthogonal to x, so $x/\|x\|\notin B$. Hence, $B\cup\{x/\|x\|\}$ is an orthogonal set properly containing B, a contradiction to the fact that B is a basis. The converse follows from Lemma 2.8.

(ii)$\Rightarrow$(iii). Suppose $x\in H$. First we show that $\sum_{b\in B}\langle x,b\rangle b$ is Cauchy, hence converges in H. Let $B=\{b_1,b_2,\dots\}$. From Corollary 2.12A we have that $\sum_{j=1}^{\infty}|\langle x,b_j\rangle|^2\leqq\|x\|^2$, so that the sequence $\langle\!\langle\sum_{j=1}^{k}|\langle x,b_j\rangle|^2\rangle\!\rangle$ $(k\in\mathbb{N})$ is Cauchy in $\mathfrak{C}$. So, for every $\varepsilon>0$, there exists $N_\varepsilon\in\mathbb{N}$ such that for all $n>m>N_\varepsilon$, $|\sum_{j=1}^{n}|\langle x,b_j\rangle|^2-\sum_{j=1}^{m}|\langle x,b_j\rangle|^2|=|\sum_{j=m}^{n}|\langle x,b_j\rangle|^2|=\sum_{j=m}^{n}|\langle x,b_j\rangle|^2<\varepsilon$. Thus, for all $n>m>N_\varepsilon$, $\|\sum_{j=1}^{n}\langle x,b_j\rangle b_j-\sum_{j=1}^{m}\langle x,b_j\rangle b_j\|^2=\|\sum_{j=m}^{n}\langle x,b_j\rangle b_j\|^2=\sum_{j=m}^{n}\|\langle x,b_j\rangle b_j\|^2=\sum_{j=m}^{n}|\langle x,b_j\rangle|^2\|b_j\|^2=\sum_{j=m}^{n}|\langle x,b_j\rangle|^2<\varepsilon$. This completes the convergence proof.

Let $y=x-\sum_{b\in B}\langle x,b\rangle b$. We shall show that $y\perp b$ for every $b\in B$, so that $y=0$. Suppose $b_0\in B$. Then $\langle y, b_0\rangle=\langle x-\sum_{b\in B}\langle x, b\rangle b, b_0\rangle=\langle x, b_0\rangle-\langle\sum_{b\in B}\langle x,b\rangle b,b_0\rangle=\langle x,b_0\rangle-\sum_{b\in B}\langle\!\langle x,b\rangle b,b_0\rangle=\langle x,b_0\rangle-\sum_{b\in B}\langle x,b\rangle\langle b,b_0\rangle=\langle x,b_0\rangle-\langle x,b_0\rangle\langle b_0,b_0\rangle=0$. The third equality follows from Theorem 2.21C because we have already shown that $\langle\!\langle\langle x,b\rangle b\rangle\!\rangle$ is a summable sequence.

(iii)$\Rightarrow$(iv). $\langle x,y\rangle=\langle\sum_{b\in B}\langle x,b\rangle b,y\rangle=\sum_{b\in B}\langle\!\langle x,b\rangle b,y\rangle=\sum_{b\in B}\langle x,b\rangle\langle b,y\rangle$.

(iv)$\Rightarrow$(v). This follows directly from the fact that $\langle x,b\rangle\langle b,x\rangle=\langle x,b\rangle\langle x,b\rangle^*=|\langle x,b\rangle|^2$.

(v)$\Rightarrow$(i). If $x\perp b$ for all $b\in B$, then $\|x\|^2=0$ by (v). Then $x=0$, so $x\notin B$. In other words, B is a maximal orthonormal set.

2.28. Examples and Projects

Consider first Example 2.3B. For every $k\in\mathbb{N}$, let $x_k=(0,0,\dots,1,0,\dots)$ with 1 in the kth place. The set $B=\{x_k|k\in\mathbb{N}\}$ is an infinite orthonormal set in H.

For Example 2.3C, for every $k\in\mathbb{N}$, let $\psi_k(t)=\sqrt{2}\sin 2\pi kt$ for all $t\in S$. Standard integral theorems establish that $\{\psi_k|k\in\mathbb{N}\}$ is an orthonormal set in H.

Chapter 3A

3A.4. Examples

A. Clearly $\mathscr{Q}$ is a quasimanual. If $x, y \in X$ then $\{x\}, \{x, y\} \in \mathscr{Q}$, violating property (ii) in Definition 3A.3.

B. Properties (ii) and (iii) in the definition of manual are easily verified. To verify (i) suppose A, B, C, D are events in $\mathfrak{M}$ with A oc B, B oc C, and C oc D. Then $\cup A \cap \cup B = \emptyset, \cup B \cap \cup C = \emptyset, \cup C \cap \cup D = \emptyset$. Further, $\mathfrak{R} = \cup A \dot{\cup} \cup B = \cup B \dot{\cup} \cup C = \cup C \dot{\cup} \cup D$. (The symbol $\dot{\cup}$ means disjoint union—the union of two disjoint sets.) From this we conclude that $\cup A \cap \cup D = \emptyset$, and $\cup A \dot{\cup} \cup D = \mathfrak{R}$. Thus, $A \cup D$ forms a countable partition of $\mathfrak{R}$.

C. It is easy to verify properties (ii) and (iii) in the definition of a manual. To verify property (i) note that if A oc B, B oc C, and C oc D, then A and C span the same subspace of H, and so do B and D. Moreover, the subspace spanned by D is orthogonal to the subspace spanned by A, so $A \cup D$ is an orthonormal basis for H.

3A.7. Project

If A op C, then there exists event B with A oc B and B oc C. If A is tested by experiment $E = A \cup B$ and A occurs, then B does not. If C is tested by experiment $F = B \cup C$ at the same time E is performed, then since no outcome in B can occur, C must occur. Similar reasoning works for the case A does not occur when tested by E.

3A.12. Example

See Figure CO3A.1.

3A.13. Project

The quasimanual $\mathscr{S}$ is not classical because outcomes in separate dichotomies are not compatible.

To verify that $\mathscr{S}$ is a manual, suppose A oc B, B oc C, and C oc D. If A has two members, then $B = \emptyset$, hence C has two members. Thus, $D = \emptyset$, which shows that D is orthogonal to A. If $A = \emptyset$, then it is certainly orthogonal to D. Finally, if A has one member, say x, then B has one member, say y. But then C has one member, and it must be x because the dichotomies are disjoint. So D has one member, and it must be y. Thus D is orthogonal to A.

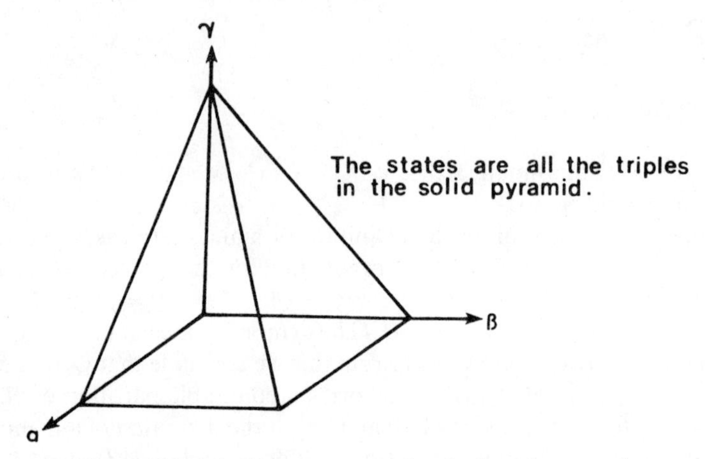

Figure CO3A.1. The state space for the bowtie manual.

3A.15. Project

Suppose ψ is a unit vector. Then for every $u \in U$, the values of ω_ψ on the outcomes u-up and u-down in E_u satisfy: $\omega_\psi(u\text{-up}) + \omega_\psi(u\text{-down}) = \| \mathrm{Proj}_{K_u} \psi \|^2 + \| \mathrm{Proj}_{[K_u^\perp]} \|^2 = \| \psi \|^2 = 1$. Further, for every vector ψ either $0 < \| \mathrm{Proj}_{K_u} \psi \|^2 < 1$ or $0 < \| \mathrm{Proj}_{[K_u^\perp]} \psi \|^2 < 1$, so ω_ψ is nonclassical.

3A.16. Project

Two classical weights on $\mathfrak{M}$ are:

$$\omega_1(r) = \omega_1(f) = 1, \quad \text{while } \omega_1(x) = 0 \text{ for } x = l, b, n;$$

and

$$\omega_2(l) = \omega_2(f) = 1, \quad \text{while } \omega_2(x) = 0 \text{ for } x = r, b, n.$$

The only unit vector ψ inducing a classical weight on $\mathfrak{M}$ is $\psi = n$.

Chapter 3B

3B.2. Lemmas

A. The task is to establish the transitivity property: if $A \leqq B$ and $B \leqq C$, then $A \leqq C$. Suppose there exist event U with $A \perp U$ and $(A \cup U)$ op B and event V with $B \perp V$ and $(B \cup V)$ op C. Then we have the situation suggested by Figure CO3B.1a, with L oc $(A \cup U)$ and K oc $(B \cup V)$. Then $(V \cup K)$

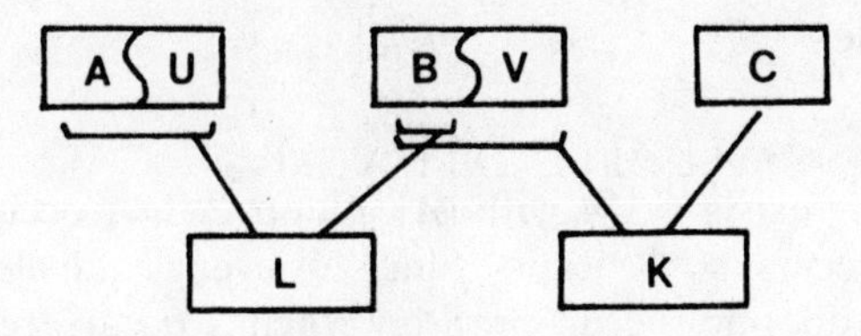

Figure CO3B.1a. oc diagram for the proof of Lemma 3B.2.

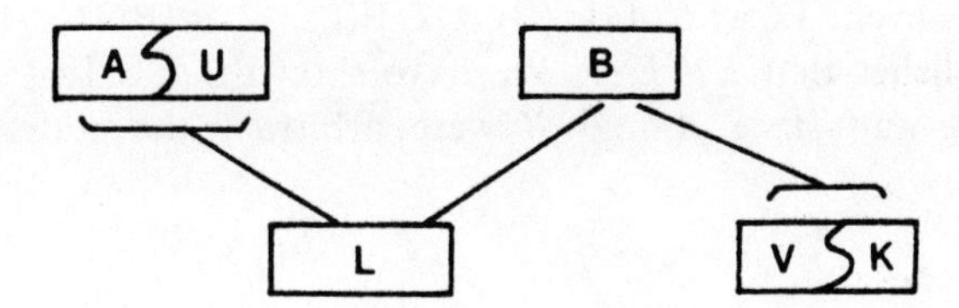

Figure CO3B.1b. α diagram for the proof of Lemma 3B.2.

oc B, so by property (i) in Definition 3A.3 we have $(V \cup K) \perp (A \cup U)$ (see Figure CO3B.1b). This implies that $K \perp (A \cup U)$, so there exists event W with $(A \cup U \cup V \cup W)$ oc K, which implies that $A \leqq C$.

B. Suppose $A \perp U$ and $(A \cup U)$ oc L, and L oc B. Then $\omega(A) + \omega(U) + \omega(L) = 1 = \omega(L) + \omega(B)$. So $\omega(A) \leqq \omega(B)$.

3B.4. Lemmas

A. Clearly, A op B implies $A \leftrightarrow B$. For the converse, suppose that $A \leqq B$ and $B \leqq A$, so that we have the situation suggested in Figure CO3B.1a with $C = A$. Then, as in the proof of Lemma 3B.2A, we have that there exists event W such that $A \cup U \cup V \cup W \cup K$ is an experiment, and all five sets in this union are disjoint from the others. Since A oc K, however, we know that $A \cup K$ is an experiment, so by property (ii) in Definition 3A.3, $U = V = W = \varnothing$. So A op B.

B. For $E, F \in \mathfrak{M}, E$ op F.

3B.5. Lemmas

A. Suppose $A \leftrightarrow C$, $B \leftrightarrow D$, and $A \leqq B$. Then $C \leqq A$ and $B \leqq D$. By transitivity we conclude that $C \leqq D$. Conversely, if $C \leqq D$, then $A \leqq C \leqq D \leqq B$.

B. Since A oc $E \backslash A$ and A oc $F \backslash A$, we have $E \backslash A$ op $F \backslash A$.

3B.7. Examples

A. The logic consists of $[\varnothing], [x], [y], [\{x, y\}]$.

B. If $A \leqq B$, there exists $C \perp A$ with $(A \cup C)$ op B. If A occurs when tested, then $A \cup C$ occurs, so B occurs. Since an event equivalent to A occurs (respectively, does not occur) precisely when A occurs (respectively, does not occur), and the same is true for B, we have established the desired result.

C. We must establish that if $A \subseteq E \in \mathfrak{M}$ and $B \subseteq F \in \mathfrak{M}$, then $A \leqq F \backslash B$ if and only if $B \leqq E \backslash A$. Suppose $A \leqq F \backslash B$. Then there exists event $C \perp A$ with $(A \cup C)$ op $F \backslash B$. Then $B \perp (A \cup C)$ by property (i) in Definition 3A.3. So there exists event W with A oc $(B \dot{\cup} C \dot{\cup} W)$ so that $(B \dot{\cup} C \dot{\cup} W)$ op $E \backslash A$. This establishes that $B \leqq E \backslash A$. We have shown that if $[A] \leqq [B]'$, then $[B] \leqq [A]'$, and since A and B were arbitrary the converse follows immediately.

3B.8. Lemma

Suppose $[A] \leqq [B]' = [E \backslash B]$ with $B \subseteq E \in \mathfrak{M}$. Then $A \leqq E \backslash B$, so there exists event K with $K \perp A$ and $(A \cup K)$ op $E \backslash B$. Then by property (i) in Definition 3A.3, $B \perp (A \dot{\cup} K)$, so $B \perp A$. The converse is easily established.

3B.11. Examples

A. The join of two subsets is their union; their meet is their intersection. The empty set is the zero, and the set X is the unit for the lattice. The tree diagram for Sub(X) where $X = \{x, y, z\}$ is in Figure CO3B.2a.

B. The join of two subspaces is the subspace of H generated by the union

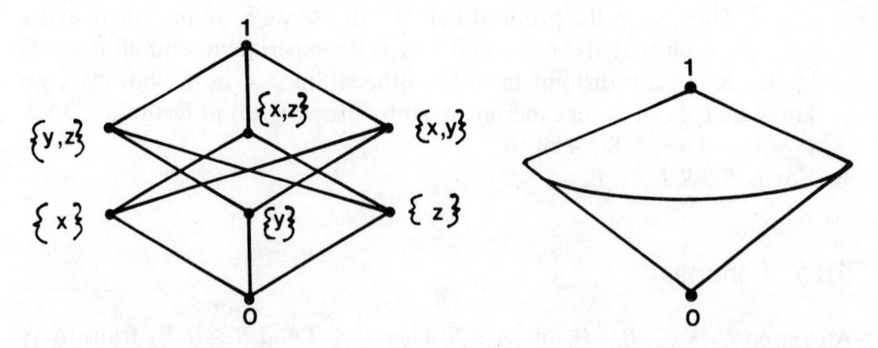

Figure CO3B.2a. The lattice Sub(x) for a set x with three members.

Figure CO3B.2b. The subspace lattice for a two-dimensional Hilbert space.

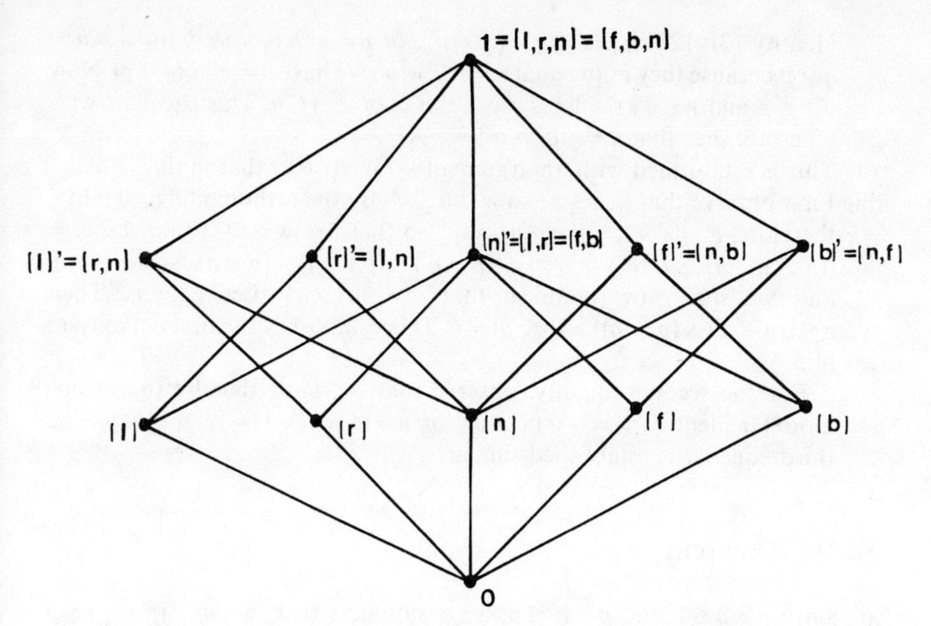

Figure CO3B.3. The operational logic for the bowtie manual.

of the two subspaces; the meet of two subspaces is their intersection. The zero for L is $\{0_H\}$, and the unit is H. The tree diagram for two-dimensional space H is in Figure CO3B.2b. There are infinitely many one-dimensional subspaces, which we have represented by a circle.

C. The tree diagram for the bow tie logic is in Figure CO3B.3.

3B.13. Theorem

We establish the first equality first. Since $p \leqq p \vee q$ and $q \leqq p \vee q$, we have $(p \vee q)' \leqq p'$ and $(p \vee q)' \leqq q'$, hence $(p \vee q)' \leqq p' \wedge q'$. Conversely, $p' \wedge q' \leqq p'$ and $p' \wedge q' \leqq q'$, so $p \leqq (p' \wedge q')'$ and $q \leqq (p' \wedge q')'$, establishing that $p \vee q \leqq (p' \wedge q')'$. Thus $p' \wedge q' \leqq (p \vee q)'$.

The second equality now follows easily. From the first equality, $(p' \vee q')' = p \wedge q$. So $(p \wedge q)' = p' \vee q'$.

3B.15. Theorem

(i) Since $u \leqq w'$ and $u \leqq v'$, we have $u \leqq v' \wedge w' = (v \vee w)' = q'$. But also $u \leqq p$, so

$$u \leqq p \wedge q'. \qquad (*)$$

Now $p \wedge q' \leqq q'$, thus $u \leqq q'$, implying that $q \leqq u'$, so by the orthomodular identity [3B.12 (iii)], $u' = q \vee (q' \wedge u')$, or $u = q' \wedge (q \vee u)$. Now $q \vee u = q \vee p$ because they both equal $u \vee v \vee w$, so we have $u = q' \wedge (q \vee p)$. Now if $r \leqq p$ and $r \leqq q'$, then $r \leqq q' \wedge p \leqq q' \wedge (q \vee p) = u$. This together with $(*)$ establishes that $u = \text{glb}\{p, q'\} = p \wedge q'$.

(ii) This is established with an argument symmetric to that in (i).

(iii) First observe that $w, u \leqq v'$, so $w \vee u \leqq v'$. By the orthomodular identity, therefore, $v' = w \vee u \vee ((w \vee u)' \wedge v')$, so that $v = (w \vee u)' \wedge (w \vee u \vee v) = ((p' \wedge q) \vee (p \wedge q'))' \wedge (p \vee q) = ((p \vee q') \wedge (p' \vee q)) \wedge (p \vee q)$. Now $v \leqq p, q$, and we shall now establish that $v = \text{glb}\{p, q\}$. Let $r \leqq p, q$. Then $r \leqq ((p \vee q') \wedge (p' \vee q)) \wedge (p \vee q) = v$. This establishes the first equality in (iii).

For the second equality, observe that $u \leqq v'$ so that by the orthomodular identity, $v' = u \vee (u' \wedge v')$, or $v = u' \wedge (u \vee v) = (p' \vee q) \wedge p$. The third equality is established similarly.

3B.16. Theorem

(i) Since $p \wedge q \leqq p$ and $p \wedge q' \leqq p$, we easily have that $(p \wedge q) \vee (p \wedge q') \leqq p$ whether or not p and q are compatible. The reverse inequality follows from Theorem 3B.15; for if $\{u, v, w\}$ is a compatibility decomposition for p and q, then $p \wedge q = v$, $p \wedge q' = u$, and $u \vee v = p$.

(ii) This follows from an argument similar to the one in (i). Conversely, if (i) and (ii) hold, we have that $\{p \wedge q, p \wedge q', q \wedge p'\}$ form a compatibility decomposition for p and q.

3B.17. Theorem

If A and B are compatible, then $\{[A \backslash B], [A \cap B], [B \backslash A]\}$ forms a compatibility decomposition for $[A]$ and $[B]$.

3B.19. Example

If $[A] \perp [B]$ in $\Pi(\mathfrak{M})$, then $A \perp B$ by Lemma 3B.8, and $[A] \vee [B] = [A \cup B]$. So $s([A] \vee [B]) = s([A \cup B]) = \omega(A \cup B) = \omega(A) + \omega(B) = s([A]) + s([B])$.

3B.20. Lemma

If $p, q \in L$ and $p \perp q$, then $s(p \vee q) = \sum_{k=1}^{n} \alpha_k s_k(p \vee q) = \sum_{k=1}^{n} \alpha_k (s_k(p) + s_k(q)) = \sum_{k=1}^{n} \alpha_k s_k(p) + \sum_{k=1}^{n} \alpha_k s_k(q) = s(p) + s(q)$. Further, $s(1_L) = \sum_{k=1}^{n} \alpha_k s_k(1_L) = \sum_{k=1}^{n} \alpha_k = 1$.

3B.23. Project

If we look into side E and see a light on the left, the firefly is lit in a chamber such that it would be seen on the right if we had looked into side A. If we see no light when we look into E, however, we do not know if the firefly is unlit or is lit in a chamber that we cannot see.

Chapter 4

4.3. Theorem

Suppose $\langle\langle x_k \rangle\rangle$ is a Cauchy sequence in $\cap \mathbb{S}$, which converges to $x \in H$. Then for all $C \in \mathbb{S}$, $\langle\langle x_k \rangle\rangle$ is a Cauchy sequence in C, and since C is closed, x must belong to C. So $x \in \cap \mathbb{S}$.

4.5. Example

For all $k \in \mathbb{N}$, $\|(1/\|x\| - 1/k)x\| < (1/\|x\|)\|x\| = 1$, so $x_k \in S$. It is easy to see that $\lim_{k \to \infty} \|x_k - x/\|x\|\| = \lim_{k \to \infty}(1/k)\|x\| = 0$.

4.7. Example

Clearly, M is a linear manifold. For each $k \in \mathbb{N}$ define $x_k = \langle\langle x_k^j \rangle\rangle$ $(j \in \mathbb{N})$, where

$$x_k^j = \begin{cases} \sqrt{1/2^j} & \text{if } j \leqq k, \\ 0 & \text{if } j > k. \end{cases}$$

Then $\langle\langle x_k \rangle\rangle$ is a Cauchy sequence in M that converges in norm to $x = \langle\langle \sqrt{1/2^j} \rangle\rangle$ $(j \in \mathbb{N})$, which is not a member of M.

4.9. Lemmas

A. Suppose K is a subspace in H. Then by Zorn's lemma there is a maximal orthonormal subset B of K contained in a basis for H. Since B is contained in a basis for H, B is countable and is a basis for K. If $x \in K$, then $x = \sum_{b \in B} \langle x, b \rangle b$ by Theorem 2.24(iii). Conversely, if $\langle\langle \sum_{j=1}^{n} \langle x, b_j \rangle b_j \rangle\rangle$ $(n \in \mathbb{N})$ is a sequence in K converging in norm to x, then $x \in K$, because K is closed.

B. Let B be a basis for K. If $y \in K$, then $\langle y, x \rangle = \langle \sum_{b \in B} \langle y, b \rangle b, x \rangle = \sum_{b \in B} \langle y, b \rangle \langle x, b \rangle = 0$.

4.10. Theorem

To show that clos (M) is a linear manifold suppose $x, y \in \text{clos}(M)$. Then there exist sequences $\langle\langle x_k \rangle\rangle$, $\langle\langle y_k \rangle\rangle$ in M that converge to x and y, respectively. By Theorem 2.21B $\langle\langle x_k + y_k \rangle\rangle$ is a sequence in M that converges to $x + y$, proving that $x + y \in \text{clos}(M)$. A similar argument shows that for $x \in \text{clos}(M)$ and $\lambda \in \mathfrak{C}$, $\lambda x \in \text{clos}(M)$.

The observation that clos (M) is a closed set completes the proof that clos (M) is a subspace in H.

4.12. Theorem

(i). It is easy to show that $\vee S$ is a linear manifold. That it is closed follows from Theorem 4.3.

Properties (ii) and (iii) follow easily from the definition of intersection.

4.13. Theorem

Property (i) is obvious, so we turn immediately to (ii). From Theorem 4.10 we know that clos (M) is a subspace in H. Since $S \subseteq M \subseteq \text{clos}(M)$, we have from Theorem 4.12(iii) that $\vee S \subseteq \text{clos}(M)$. On the other hand, $\vee S$ is a linear manifold, which implies that every finite linear combination of members of S is a member of $\vee S$. Thus, $M \subseteq \vee S$. Since $\vee S$ is closed, clos $(M) \subseteq \vee S$ in light of the remark following Definition 4.4.

4.16. Lemmas

A. If S is a basis for $\vee S$, it is a countable orthonormal set. Conversely, suppose $x \in \vee S$. Then x can be written as a linear combination $x = \sum_{k=1}^{\infty} \lambda_k x_k$ with $x_k \in S$ for all $k \in \mathbb{N}$. Now for each $k \in \mathbb{N}$, $\langle x, x_k \rangle = \sum_{j=1}^{\infty} \lambda_j \langle x_k, x_j \rangle = \lambda_k$. So if $x \perp s$ for all $s \in S$, $x = 0$. This proves that S is a maximal orthonormal set in $\vee S$.

B. $A \subseteq (B \backslash A)^{\perp}$, so $\vee A \subseteq (B \backslash A)^{\perp}$. Conversely, if $x \in (B \backslash A)^{\perp}$, then $x = \sum_{b \in B \backslash A} \langle x, b \rangle b + \sum_{b \in B} \langle x, b \rangle b = \sum_{b \in A} \langle x, b \rangle b \in \vee A$.

4.17. Theorem

(i) If $x \in S \cap S^{\perp}$, then $x \perp x$, which implies that $x = 0$.

(ii) If $x, y \in S^{\perp}$, and $\lambda \in \mathfrak{C}$, then for all $z \in S$, $\langle x + y, z \rangle = \langle x, z \rangle + \langle y, z \rangle = 0 + 0 = 0$, and $\langle \lambda x, z \rangle = \lambda \langle x, z \rangle = 0$. Thus, $S^{\perp}$ is a linear manifold. Suppose $\langle\langle x_k \rangle\rangle$ is a sequence in $S^{\perp}$ that converges in norm to $x \in H$.

Then for all $z \in S$, $\langle x, z\rangle = \lim_{k\to\infty} \langle x_k, z\rangle$ by Theorem 2.21B(i). Since $\langle x_k, z\rangle = 0$ for all $k \in \mathbb{N}$, we have that $\langle x, z\rangle = 0$. So $x \in S^\perp$, which shows that $S^\perp$ is closed.

(iii) Suppose $x \in T^\perp$. For every $s \in S$ we have $s \in T$, so that $x \perp s$. Hence, $x \in S^\perp$.

(iv) Suppose $x \in S$. If $t \in S^\perp$, then $x \perp t$. Thus, $x \in (S^\perp)^\perp$.

(v) From (iv) we know that $S \subseteq S^{\perp\perp}$. If $S \neq S^{\perp\perp}$, then there exists nonzero vector $x \in S^{\perp\perp}$ with $x \notin S$. Let B be a basis for S. Let $y = x - \sum_{b\in B} \langle x, b\rangle b$. Then $y \in S^{\perp\perp}$, and since $x \notin S$, $y \neq 0$ by Lemma 4.9A. Using the same technique as in the proof that (ii) implies (iii) in Theorem 2.24, we can show that $y \perp b$ for all $b \in B$. Then by Lemma 4.9B, $y \in S^\perp$. So $y \in S^\perp \cap S^{\perp\perp}$, contradicting (i).

(vi) Let us write $\vee\, \mathbb{S}$ for $\vee_{S\in\mathbb{S}} S$, and write $\cap\, \mathbb{S}$ for $\bigcap_{S\in\mathbb{S}} S$.

We prove the first equality first. For all $S \in \mathbb{S}$ we have $S \subseteq \vee\, \mathbb{S}$, so $(\vee\, \mathbb{S})^\perp \subseteq S^\perp$. Conversely, suppose $x \in \bigcap_{S\in\mathbb{S}} S^\perp$. If $y \in \vee\, \mathbb{S}$, then y is a linear combination $y = \sum_{k=1}^{\infty} \lambda_k y_k$ with $y_k \in \cup\, \mathbb{S}$ for all $k \in \mathbb{N}$. Then $x \perp y_k$ for all $k \in \mathbb{N}$, hence $x \perp y$.

For the second equality, we use the first equality and (v) to obtain $(\vee_{S\in\mathbb{S}} S^\perp)^\perp = \bigcap_{S\in\mathbb{S}} S^{\perp\perp} = \bigcap_{S\in\mathbb{S}} S$. Using (v) again gives the second equality.

(vii) If $x \in \vee\, S$, $x = \sum_{k=1}^{\infty} \lambda_k x_k$ with $x_k \in S$ for all $k \in \mathbb{N}$. But $\lambda_k = \langle x, x_k\rangle$ for all $k \in \mathbb{N}$ by Theorem 2.21D, so if $x \in S^\perp$, $x = 0$.

4.18. Theorem

To establish the orthomodular identity in $\mathbb{L}$, suppose $J, K \in \mathbb{L}$ with $J \subseteq K$. We wish to show that $K = J \vee (J^\perp \cap K)$. By Theorem 4.8 there exist bases B and B_0 for J and K, respectively, with $B \subseteq B_0$.

First, suppose $x \in K$. Then $x = \sum_{b\in B_0} \langle x, b\rangle b = \sum_{b\in B} \langle x, b\rangle b + \sum_{b\in B_0\setminus B} \langle x, b\rangle b$. Since $B_0 \setminus B \subseteq J^\perp$, the second term belongs to $J^\perp \cap K$. Since the first term belongs to J, we have that x is a linear combination of vectors in $J \cup (J^\perp \cap K)$, so $x \in J \vee (J^\perp \cap K)$. This establishes that $K \subseteq J \vee (J^\perp \cap K)$.

On the other hand, observe that since both J and $J^\perp \cap K$ are subspaces of K, so is their join.

4.19. Lemma

A. Suppose $A \operatorname{oc} C$. Then $A \cup C$ is a basis for H. Now $C^\perp = ((A \cup C)\setminus A)^\perp$, which equals $\vee\, A$ by 4.16B.

Now suppose $C^\perp = \vee\, A$. Then $A \subseteq \vee\, A = C^\perp$, so $A \perp C$. Further, $(A \cup C)^\perp = A^\perp \cap C^\perp = A^\perp \cap \vee\, A = \{0\}$, so that $A \cup C$ is a maximal orthonormal set. This proves that $A \operatorname{oc} C$.

B. Suppose $A \operatorname{op} B$. Then there exists event C with $A \operatorname{oc} C$ and $B \operatorname{oc} C$. By part A, we conclude that $\vee\, A = C^\perp = \vee\, B$.

Suppose that $\vee A = \vee B$. Let $C = \vee A$. Then $C^{\perp} = \vee A = \vee B$, so by Part A, $A \operatorname{oc} C$ and $B \operatorname{oc} C$. Hence $A \operatorname{op} B$.

4.20. Theorem

(i) Lemma 4.19 establishes the equivalence of (a) and (b). The equivalence of (a) and (c) follows from Lemma 3B.4A

(ii) Suppose $B \subseteq E \in \mathscr{F}(H)$. Then $[A] \perp [B]$ if and only if there exists event C with $\vee (A \dot{\cup} C) = \vee (E \backslash B)$, which is equivalent to the statement $A \perp B$.

4.22. Theorem

(i) Clearly, $K + L$ is a linear manifold. Our proof will be complete when we show that it is closed. Suppose $\langle\!\langle x_j + y_j \rangle\!\rangle$ is a Cauchy sequence in H with $x_j \in K$ and $y_j \in L$ for all $j \in \mathbb{N}$. Then for all $\varepsilon > 0$, there exists N_ε such that for all $m, n > N_\varepsilon$, $\| x_m - x_n \|^2 + \| y_n - y_m \|^2 = \| x_m + y_m - (x_n + y_n) \|^2 < \varepsilon$, the equality following from the Pythagorean theorem and the fact that $K \perp L$. This proves that $\langle\!\langle x_j \rangle\!\rangle$ and $\langle\!\langle y_j \rangle\!\rangle$ are Cauchy sequences, and since K and L are closed, they must converge, respectively, to $x \in K$ and $y \in L$. By Theorem 2.21B(ii) we conclude that $\lim_{j \to \infty} \langle x_j + y_j \rangle = x + y \in K + L$.

(ii) Since $K + L$ is a subspace, and $K \cup L \subseteq K + L$, we have $K \vee L \subseteq K + L$ by Theorem 4.12. Conversely, $K \vee L$ is a linear manifold containing both K and L, so it clearly contains $K + L$.

4.23. The Finite Projection Theorem

Since $K \perp K^{\perp}$ and both K and $K^{\perp}$ are subspaces, $K \vee K^{\perp}$ is a subspace in H. If it is a proper subspace, we can find a basis B for it, and a basis B_0 for H with $B \subset B_0$ ($B \neq B_0$). Then there is a nonzero vector $x \in B_0 \backslash B$, so that $0 \neq x \in (\vee B)^{\perp} = (K \vee K^{\perp})^{\perp} = K^{\perp} \cap K = \{0\}$, a contradiction. Thus, $K + K^{\perp} = K \vee K^{\perp} = H$ so every vector $x \in H$, can be written as $x = x_1 + x_2$ with $x_1 \in K$ and $x_2 \in K^{\perp}$. If also $x = y_1 + y_2$ with $y_1 \in K$ and $y_2 \in K^{\perp}$, then $0 = \| 0 \|^2 = \| x_1 - y_1 + x_2 - y_2 \|^2 = \| x_1 - y_1 \|^2 + \| x_2 - y_2 \|^2$ by the Pythagorean theorem.

4.24. Theorem

If $x = \sum_{j=1}^{\infty} x_j$ with $x_j \in K_j$ for all $j \in \mathbb{N}$, then $x_j \in \vee \mathbb{K}$ for all $j \in \mathbb{N}$, so $x \in \vee \mathbb{K}$. Thus $+\mathbb{K} \subseteq \vee \mathbb{K}$.

To prove the converse, suppose $z \in \vee \mathbb{K}$. By the finite projection theorem

(Theorem 4.23), for every $j\in\mathbb{N}$ there exist $x_j\in K_j$ and $y_j\in K_j^\perp$ with $z=x_j+y_j$. We shall show that $\langle\!\langle x_j\rangle\!\rangle$ is summable and that its sum is z.

For $x_j\neq 0$, $\langle z, x_j/\|x_j\|\rangle=(1/\|x_j\|)\langle z,x_j\rangle=(1/\|x_j\|)\langle x_j+y_j,x_j\rangle=(1/\|x_j\|)(\langle x_j,x_j\rangle+\langle y_j,x_j\rangle)=\|x_j\|^2/\|x_j\|=\|x_j\|$. The fourth equality attains because $y_j\perp x_j$. From Bessel's inequality (Theorem 2.11) $\sum_{j=1}^\infty\|x_j\|^2=\sum_{j=1(0\neq x_j)}^\infty|\langle z,x_j/\|x_j\|\rangle|^2\leqq\|z\|^2$. From Theorem 2.21E, therefore, $\langle\!\langle x_j\rangle\!\rangle$ is summable, say to $w\in H$. It remains to show that $z-w=0$. We shall accomplish this by showing $z-w\in\vee\mathbb{K}\cap(\vee\mathbb{K})^\perp$.

Since w is a sum of vectors from $\cup\mathbb{K}$, we have $w\in\vee\mathbb{K}$, hence $z-w\in\vee\mathbb{K}$. On the other hand, if $k\in\mathbb{N}$ and $v\in K_k$, then $\langle z-w,v\rangle=\langle x_k+y_k-\sum_{j=1}^\infty x_j,v\rangle=\langle x_k,v\rangle+\langle y_k,v\rangle-\sum_{j=1}^\infty\langle x_j,v\rangle=\langle x_k,v\rangle-\langle x_k,v\rangle=0$, which shows that $z-w\in K_k^\perp$. We conclude that $z-w\in\bigcap_{K_j\in\mathbb{K}}K_j^\perp=(\vee\mathbb{K})^\perp$.

4.26. The General Projection Theorem

From Theorem 4.24 we know that if $x\in\vee\mathbb{K}$, we can write x as a sum $x=\sum_{j=1}^\infty x_j$ with $x_j\in K_j$ for all $j\in\mathbb{N}$. Then $\|x\|^2=\sum_{j=1}^\infty\|x_j\|^2$ by Theorem 2.21E.

To prove uniqueness, suppose that also $x=\sum_{j=1}^\infty y_j$. Then $0=\sum_{j=1}^\infty x_j-y_j$, so that $0=\sum_{j=1}^\infty\|x_j-y_j\|^2$. Hence, $x_j=y_j$ for all $j\in\mathbb{N}$.

Chapter 5A

5A.3. Lemma

A. $A(0)=A(0x)=0A(x)=0$ for any $x\in M_1$.
B. This is what it means to say that $\|A\|$ is the *greatest lower bound* of all t for which $\|Ax\|\leqq t\|x\|$ for all $x\in M_1$.
C. For all $t\in\mathfrak{R}$, if $x\in M_1$ and $\|Ax\|\leqq t\|x\|$, then, since $0\leqq\|Ax\|$, we have $t\geqq 0$. Thus, $\|A\|\geqq 0$.

If A is the zero map, then for all $x\in M_1$, $\|Ax\|\leqq 0\|x\|$. So $\|A\|\leqq 0$ from B above so $\|A\|=0$. Conversely, if $\|A\|=0$, then $\|Ax\|\leqq\|A\|\,\|x\|\leqq 0$ for all $x\in M_1$, which implies that A is the zero map.

5A.5. Theorem

(i) implies (ii). Let $r=1$. For $\|x\|\leqq 1$, $\|Ax\|\leqq\|A\|\,\|x\|\leqq\|A\|$.

(ii) implies (i). Suppose that $r,t>0$ and for all $y\in M_1$ if $\|y\|\leqq r$, then $\|Ay\|\leqq t$. Then for all $x\in M_1$ with $x\neq 0$. $\|(1/\|x\|)rx\|\leqq r$, so $(r/\|x\|)\|Ax\|=\|A((1/\|x\|)rx)\|\leqq t$, which implies that $\|Ax\|\leqq\|x\|t/r$. Since this last inequality obviously holds even if $x=0$, the proof is complete.

(i) implies (iii). Consider $r>0$. If $y \in B^r_{M_1}$, then $\| Ay \| \leqq \| A \| \| y \| \leqq \| A \| r$.

(iii) implies (ii). This is obvious.

(iii) implies (iv). This is obvious.

(iv) implies (iii). If (iii) is false, then we can find $r>0$ with $A[B^r_{M_1}]$ unbounded. Thus for every $k \in \mathbb{N}$, there exists $y_k \neq 0$ with $\| y_k \| < r$ and $\| Ay_k \| > kr$. Let $x_k = y_k / \| y_k \|$ for every k. Then $\| Ax_k \| = \| Ay_k \| / \| y_k \| > k$ for every k, and the existence of the sequence $\langle\langle x_k \rangle\rangle$ shows that (iv) is also false.

5A.6. Examples

A. Let $\mathscr{A} = [[a_{ij}]]_{i=1,m,\, j=1,n}$ be the matrix representation of A with respect to the standard bases for $\mathbb{C}^n$ and $\mathbb{C}^m$. We will show that $\| Ax \| \leqq t \| x \|$ for $x \in B^1_{\mathbb{C}^n}$, where t is a function of the entries in A. This will complete the proof in light of Theorem 5A.5(ii).

For $x = (x_1, \ldots, x_n) \in B^1_{\mathbb{C}^n}$, $\mathscr{A}x = (\sum_{j=1}^n a_{1j}x_j, \ldots, \sum_{j=1}^n a_{mj}x_j)$. Then $\| Ax \|^2 \leqq \sum_{k=1,m,\, j=1,n} |a_{kj}x_j|^2 \leqq \sum_{k=1,m,\, j=1,n} |a_{kj}|^2$, because $|x_j|^2 \leqq 1$ for $j = 1, \ldots, n$.

B. Let $x = x_1 + x_2$, $y = y_1 + y_2$. Then $(x_1 + y_1) + (x_2 + y_2)$ is one way of writing vector $x + y$ as a sum of $x_1 + y_1 \in K$ and $x_2 + y_2 \in K^\perp$. So it must be the only way. Thus, $P_K(x+y) = x_1 + y_1 = P_K x + P_K y$. A similar argument establishes that $P_K(\lambda x) = \lambda P_K x$ for all $\lambda \in \mathbb{C}$.

Further, for all $x \in H$, if $x = x_1 + x_2$ as above, then $\| P_K x \|^2 = \| x_1 \|^2 \leqq \| x_1 \|^2 + \| x_2 \|^2 = \| x_1 + x_2 \|^2 = \| x \|^2$. The second equality follows from the Pythagorean theorem because $x_1 \perp x_2$. So $\| P_K x \| \leqq \| x \|$ for all $x \in H$, which implies that $\| P \| \leqq 1$.

C. Figure CO5A.1 shows a sketch of the graph of ψ_n, where $n > 2/(b-a)$.

Then $\| \psi_n \|^2 = \int_a^b |\psi_n|^2 \, d\mu = \int_a^{a+1/n} n^2(t-a)^2 \, dt + \int_{a+1/n}^{a+2/n} (2 - n(t-a))^2 \, dt < 2/n$ by routine integration.

Observe next that

$$\psi_n'(t) = \begin{cases} n & \text{if } a \leqq t \leqq a + 1/n, \\ -n & \text{if } a + 1/n \leqq t \leqq a + 2/n, \\ 0 & \text{if } a + 2/n \leqq t \leqq b. \end{cases}$$

Then $\| i\psi_n' \|^2 = \int_a^b |\psi_n'|^2 \, d\mu = 2n$, again, by routine integration.

D. $\| Q\psi \|^2 = \int_a^b |Q\psi|^2 \, d\mu \leqq b^2 \int_a^b |\psi|^2 \, d\mu = b^2 \| \psi \|^2$.

5A.8. Theorem

Suppose A is bounded and is not the (obviously continuous) zero map. Consider sequence $\langle\langle x_n \rangle\rangle$ in M_1 converging in norm to $x \in M_1$. Then for every $\varepsilon > 0$, there exists $N \in \mathbb{N}$ such that for all $n > N$, $\| x_n - x \| < \varepsilon / \| A \|$. Thus for $n > N$, $\| Ax_n - Ax \| = \| A(x_n - x) \| \leqq \| A \| \| x_n - x \| \leqq \| A \| (\varepsilon / \| A \|) = \varepsilon$, which shows that $\langle\langle Ax_n \rangle\rangle$ converges in norm to Ax.

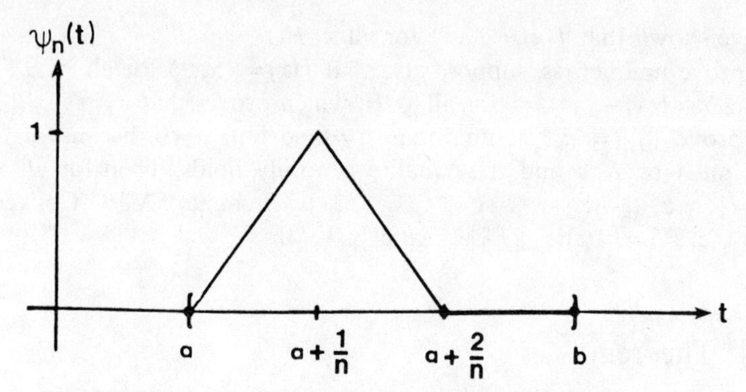

Figure CO5A.1. The graph of a function used in Example 5A.6c.

Conversely, if A is not bounded, by Theorem 5A.5(ii) for every $n \in \mathbb{N}$ there exists vector $x_n \in B_{B_1}^{1/n}$ with $\| Ax_n \| > n$. Then $\langle\!\langle x_n \rangle\!\rangle$ converges in norm to 0, but $\langle\!\langle Ax_n \rangle\!\rangle$ does not converge in norm to $A0 = 0$.

5A.11. Theorem (The Riesz Representation Theorem)

In this proof we revert to the notation $f(x)$ instead of fx to help the reader wade through some long computations.

We prove existence of z first. If $f = 0$, let $z = 0$. Now suppose $f \neq 0$. Let $K = \{x \in H \mid f(x) = 0\}$. Clearly, K is a linear manifold. It is also closed. (Those with some knowledge of topology might recognize that $K = f^{\leftarrow}\{0\}$ is closed because it is the inverse image, under the continuous function f, of a closed set in $\mathfrak{R}$. For those without such knowledge we present an alternative proof that K is closed.) Suppose $\langle\!\langle x_n \rangle\!\rangle$ is a sequence in K converging in norm to $x \in H$. Since $\langle\!\langle x - x_n \rangle\!\rangle$ converges in norm to 0 and f is continuous, we know that $\langle\!\langle f(x - x_n) \rangle\!\rangle$ converges in norm to $f(0) = 0$. But for all $n \in \mathbb{N}$, $f(x_n) = 0$, so $\langle\!\langle f(x - x_n) \rangle\!\rangle$ is a constant sequence: $\langle\!\langle f(x - x_n) \rangle\!\rangle = \langle\!\langle f(x) - f(x_n) \rangle\!\rangle = \langle\!\langle f(x) \rangle\!\rangle$. So $f(x) = 0$, which means $x \in K$.

Now since $f \neq 0$, $K \neq H$, so $K^{\perp} \neq \{0\}$. Thus, there is a nonzero unit vector $z_0 \in K^{\perp}$. Let $z = (f(z_0)^*/\langle z_0, z_0 \rangle) z_0$. Then for all $x \in H$, $f(z_0)x - f(x)z_0 \in K$, which can be verified easily by computing its image under f; hence it is orthogonal to z_0. This will give us the final equality in the following string:

$$\begin{aligned}\langle z, x \rangle &= \frac{f(z_0)^*}{\langle z_0, z_0 \rangle} \langle z_0, x \rangle = \frac{1}{\langle z_0, z_0 \rangle} \langle z_0, f(z_0)x \rangle \\ &= \frac{1}{\langle z_0, z_0 \rangle} \langle z_0, f(z_0)x - f(x)z_0 + f(x)z_0 \rangle \\ &= \frac{1}{\langle z_0, z_0 \rangle} \langle z_0, f(z_0)x - f(x)z_0 \rangle + \frac{1}{\langle z_0, z_0 \rangle} \langle z_0, f(x)z_0 \rangle = f(x)^*.\end{aligned}$$

We have shown that $f(x) = \langle x, z \rangle$ for all $x \in H$.

To prove uniqueness, suppose also that $f(x) = \langle x, z' \rangle$ for all $x \in H$. Then $\langle x, z - z' \rangle = f(x) - f(x) = 0$ for all $x \in H$, which proves that $z = z'$.

To prove $\| f \| = \| z \|$ assume that $f \neq 0$ so that $z \neq 0$, because if $f = 0$, then z must be zero and the equality certainly holds. Then for all $x \in H$, $|f(x)| = |\langle x, z \rangle| \leqq \| z \| \| x \|$, so $\| f \| \leqq \| z \|$ by Lemma 5A.3B. Conversely, $\| z \|^2 = |\langle z, z \rangle| = |f(z)| \leqq \| f \| \| z \|$, and $\| z \| \neq 0$.

5A.14. Theorem

Suppose $A \in \mathbb{B}(H)$. For each $x \in H$ consider function $f_A^x : H \to \mathbb{C}$ defined for all $y \in H$ by $f_A^x(y) = \langle Ay, x \rangle$.

We show first that $f_A^x \in \hat{H}$. It is easy to check that it is a linear map. It is bounded because for all $y \in H$, $|f_A^x(y)| = |\langle Ay, x \rangle| \leqq \| Ay \| \| x \| \leqq \| A \| \| y \| \| x \| = \| A \| \| x \| \| y \|$.

Now by the Riesz representation theorem there exists unique $z \in H$ such that for all $y \in H$, $\langle Ay, x \rangle = f_A^x(y) = \langle y, z \rangle$. We define $A^\# x = z$.Then $\langle x, Ay \rangle = f_A^x(y)^* = \langle y, A^\# x \rangle^* = \langle A^\# x, y \rangle$.

To establish that $A^\#$ is linear, suppose $x_1, x_2, y \in H$. Then $\langle A^\#(x_1 + x_2), y \rangle = \langle x_1 + x_2, Ay \rangle = \langle x_1, Ay \rangle + \langle x_2, Ay \rangle = \langle A^\# x_1, y \rangle + \langle A^\# x_2, y \rangle = \langle A^\# x_1 + A^\# x_2, y \rangle$. Since this establishes that $\langle A^\#(x_1 + x_2) - (A^\# x_1 + A^\# x_2), y \rangle = 0$ for all y, we see that $A^\#$ is additive. It is just as easy to show that $A^\#(\lambda x) = \lambda A^\#(x)$ for all $x \in H$ and all $\lambda \in \mathbb{C}$.

Finally, $\| A^\# x \|^2 = |\langle A^\# x, A^\# x \rangle| = |\langle AA^\# x, x \rangle| \leqq \| AA^\# x \| \| x \| \leqq \| A \| \| A^\# x \| \| x \|$ for all $x \in H$. So if $x \neq 0$, $\| A^\# x \| \leqq \| A \| \| x \|$, hence $A^\#$ is a bounded linear map.

5A.15. Theorem

(i) $\langle x, A^\# y \rangle = \langle A^\# y, x \rangle^* = \langle y, Ax \rangle^* = \langle Ax, y \rangle$.

(ii) For all $x, y \in H$, $\langle x, (A_1 A_2)^\# y \rangle = \langle (A_1 A_2) x, y \rangle = \langle A_2 x, A_1^\# y \rangle = \langle x, A_2^\# A_1^\# y \rangle$, the first equality following from (i). Thus, for all $y \in H$, $\langle x, (A_1 A_2)^\# y - A_2^\# A_1^\# y \rangle = 0$ for all $x \in H$, which implies that $(A_1 A_2)^\# = A_2^\# A_1^\#$.

The remaining properties are established similarly.

5A.17. Examples

Example 5A.6B. If $x, y \in H$ and $x = x_1 + x_2$, $y = y_1 + y_2$ with $x_1, x_2 \in K$ and $y_1, y_2 \in K^\perp$, then $\langle P_K x, y \rangle = \langle x_1, y_1 + y_2 \rangle = \langle x_1, y_1 \rangle + \langle x_1, y_2 \rangle = \langle x_1, y_1 \rangle = \langle x_1, y_1 \rangle + \langle x_2, y_1 \rangle = \langle x_1 + x_2, y_1 \rangle = \langle x, P_K y \rangle$.

Example 5A.6C. Recall the rule for differentiating the product of two functions. We use it in reverse to antidifferentiate as follows. For $\psi_1, \psi_2 \in M$, $\langle D\psi_1, \psi_2\rangle - \langle \psi_1, D\psi_2\rangle = \langle i\psi_1', \psi_2\rangle - \langle \psi_1, i\psi_2'\rangle = i\int_a^b \psi_1' \psi_2^* \mathrm{d}\mu + i\int_a^b \psi_1 \psi_2'^* \mathrm{d}\mu = i(\psi_1\psi_2^*(b) - \psi_1\psi_2^*(a)) = 0$.

Example 5A.6D. For $\psi_1, \psi_2 \in M$, $\langle Q\psi_1, \psi_2\rangle - \langle \psi_1, Q\psi_2\rangle = \int_a^b t\psi_1(t)\psi_2^*(t)\mathrm{d}t - \int_a^b \psi_1(t)(t\psi_2(t))^* \mathrm{d}t = 0$, because t is real, so that $t = t^*$.

Chapter 5B

5B.3. Theorem

If $x \in K^\perp$ and $y \in K$, then $\langle A^\# x, y\rangle = \langle x, Ay\rangle = 0$, because $Ay \in K$. Thus, $A^\# x \in K^\perp$.

5B.4. Theorem

$\langle Ax, x\rangle^* = \langle x, Ax\rangle^* = \langle Ax, x\rangle$. A complex number equal to its conjugate is necessarily a real number.

5B.5. Theorem

If A_1A_2 is Hermitian, $A_1A_2 = (A_1A_2)^\# = A_2^\# A_1^\# = A_2A_1$.

If A_1 and A_2 commute, then $(A_1A_2)^\# = (A_2A_1)^\# = A_1^\# A_2^\# = A_1A_2$.

5B.7. Theorem

If $P \in \mathbb{P}(H)$, $P^2 = PP^\# PP^\# = P(PP^\# P)^\# = P(PP)^\# = P(P^\# P^\#) = (PP^\#)P^\# = PP^\# = P$. This shows that P is idempotent, and it is very easy to check that $PP^\#$ is Hermitian.

If, conversely, P satisfies (i) and (ii), then $PP^\# = P^2 = P$.

5B.8. Theorem

(i) If $x = Py$ for $y \in H$, then $Px = P^2y = Py = x$.
(ii) Let $L = \{x \in H \mid Px = x\}$. From (i), $K \subseteq L$. It is obvious that $L \subseteq K$.
(iii) It is easy to show that K is a linear manifold. If $\langle\langle x_j \rangle\rangle$ is a sequence in K converging to $x \in H$, then $\langle\langle Px_j \rangle\rangle$ converges in norm to Px by Theorem 5A.8. But $\langle\langle Px_j \rangle\rangle = \langle\langle x_j \rangle\rangle$ for all $j \in \mathbb{N}$, so $\langle\langle Px_j \rangle\rangle = \langle\langle x_j \rangle\rangle$ converges in norm to x. So $Px = x$, thus $x \in K$.

(iv) It is easy to show that $I-P$ is Hermitian and idempotent. From property (ii), then, it sufficies to show that $I-P$ is the identity on $K^{\perp}$. If $y \in K^{\perp}$, then for all $x \in H$, we have that $x = x_1 + x_2$ with $x_1 \in K$ and $x_2 \in K^{\perp}$. Then $\langle (I-P)y, x\rangle = \langle y - Py, x_1 + x_2\rangle = \langle y, x_1\rangle - \langle Py, x_1\rangle + \langle y, x_2\rangle - \langle Py, x_2\rangle = -\langle Py, x_1\rangle + \langle y, x_2\rangle = -\langle y, Px_1\rangle + \langle y, x_2\rangle = \langle y, x_2\rangle$. At the same time $\langle y, x\rangle = \langle y, x_1 + x_2\rangle = \langle y, x_1\rangle + \langle y, x_2\rangle = \langle y, x_2\rangle$. So $(I-P)y = y$.

(v) First, $Px = Px_1 + Px_2 = Px_1 = x_1$. The reason $Px_2 = 0$ is $x_2 \in K^{\perp}$ and $Px_2 \in K$, so that $0 = \langle x_2, Px_2\rangle = \langle x_2, P^2x_2\rangle = \langle Px_2, Px_2\rangle$. It is now easy to show that $(I-P)x = x_2$.

(vi) From (v) we have $x = Px + (I-P)x$. From the Pythagorean theorem, then, $\|x\|^2 = \|Px\|^2 + \|x - Px\|^2$. From the hypothesis, we then have $\|x - Px\|^2 = 0$.

(vii) $\|Px\|^2 = \langle Px, Px\rangle = \langle P^2x, x\rangle = \langle Px, x\rangle$.

(viii) For all $x \in H$, $\|Px\|^2 = \langle Px, x\rangle = |\langle Px, x\rangle| \leqq \|Px\| \; \|x\|$. So $\|Px\| \leqq \|x\|$, which implies that $\|P\| \leqq 1$. Conversely, if $x \in K$, then $\|Px\| = \|x\|$, so that $\|P\| \geqq 1$.

5B.9. Theorem

(i) implies (ii). Suppose $K \perp L$ and $x \in H$. Then $Qx \in L \subseteq K^{\perp}$. So by (iv) and (ii) of Theorem 5B.8, $Qx - PQx = (I-P)Qx = Qx$. So $PQx = 0$.

(ii) implies (i). Suppose $PQ = 0$ and $x \in K$. Then if $y \in L$, $\langle x, y\rangle = \langle Px, Qy\rangle = \langle x, PQy\rangle = \langle x, 0\rangle = 0$. So $K \perp L$.

We have established that (i) is equivalent to (ii), and a similar argument will establish the equivalence of (i) and (iii).

(ii) implies (iv). If $x \in L$, $Qx = x$. Thus, $Px = PQx = 0$.

(iv) implies (ii). If $x \in H$, $Qx \in L$. Thus, $PQx = 0$.

This proves that (ii) is equivalent to (iv), and a similar proof establishes that (iii) is equivalent to (v).

5B.12. Theorem

For all $x \in H$, $\sum_{k=1}^{\infty} \|A_k x\| \leqq \sum_{k=1}^{\infty} \|A_k\| \; \|x\| = \|x\| \sum_{k=1}^{\infty} \|A_k\| < \infty$. The proof is completed by now employing Theorem 2.21E.

5B.14. Theorem

If $x \in H$, then $(\sum_{k=1}^{\infty} A_k)Bx = \sum_{k=1}^{\infty} A_k Bx$, which proves the first equality. Now for each $x \in H$, $\langle\langle \sum_{k=1}^{n} A_k x \rangle\rangle$ converges in norm to Ax. Then since B is continuous by Theorem 5A.8, we have that $B(\sum_{k=1}^{\infty} A_k x)$ is the limit of the sequence $\langle\langle B \sum_{k=1}^{n} A_k x\rangle\rangle = \langle\langle \sum_{k=1}^{n} BA_k x\rangle\rangle$, which converges in norm to $\sum_{k=1}^{\infty} BA_k x$.

5B.15. Theorem

Suppose $P_k \perp P_j$ for all $k, j \in \mathbb{N}$ $(k \neq j)$. Then $P^2 = (\sum_{k=1}^{\infty} P_k)(\sum_{j=1}^{\infty} P_j) = \sum_{k=1}^{\infty} P_k(\sum_{j=1}^{\infty} P_j) = \sum_{k=1}^{\infty} P_k^2 = \sum_{k=1}^{\infty} P_k = P$. The orthogonality was used for the third equality. Further, if $x, y \in H$, then $\langle Px, y \rangle = \langle \sum_{k=1}^{\infty} P_k x, y \rangle = \sum_{k=1}^{\infty} \langle P_k x, y \rangle = \sum_{k=1}^{\infty} \langle x, P_k y \rangle = \langle x, \sum_{k=1}^{\infty} P_k y \rangle = \langle x, Py \rangle$. So P is idempotent and Hermitian.

Suppose, conversely, that P is a projection. Then if $x \in \text{image}(P_k)$ for some $k \in \mathbb{N}$, we have $\| x \|^2 \geqq \| Px \|^2 = \langle Px, Px \rangle = \langle Px, x \rangle = \sum_{j=1}^{\infty} \langle P_j x, x \rangle = \sum_{j=1}^{\infty} \| P_j x \|^2 \geqq \| P_k x \|^2 = \| x \|^2$. The first inequality follows because $\| P \| = 1$ by Theorem 5B.8(viii). The fourth equality follows from Theorem 5B.8(vii). The last equality follows from Theorem 5B.8(ii). We conclude from this calculation that $\sum_{j=1}^{\infty} \| P_j x \|^2 = \| P_k x \|^2$, so that $P_j x = 0$ for all $j \in \mathbb{N}$ with $j \neq k$. Thus, $P_j[P_k[H]] = \{0\}$ for all $j, k \in \mathbb{N}$ with $j \neq k$. By Theorem 5B.9, we conclude that $\langle\!\langle P_k \rangle\!\rangle$ is a pairwise orthogonal sequence.

5B.16. Theorem

We establish the first equality first. If $x \in P[H]$, then for some $y \in H$, $x = Py = \sum_{j=1}^{\infty} P_j y$, and $P_j y \in K_j$ for all $j \in \mathbb{N}$. Conversely, if $x = \sum_{j=1}^{\infty} y_j$ with $y_j \in K_j$ for all $j \in \mathbb{N}$, then $x = \sum_{j=1}^{\infty} P_j y_j$, so $Px = \sum_{j=1}^{\infty} (P_j^2) y_j = \sum_{j=1}^{\infty} P_j y_j = x$. So $x \in P[H]$.

The second equality in the conclusion of the theorem follows from Theorem 4.24.

5B.18. Theorems

A. As we remarked after Definition 5B.10, the map $P \to P[H]$ is an orthogonality-preserving bijection. The remaining assertions follow from the definitions of join and meet in $\mathbb{P}(H)$.
B. The map $P \to P[H]$ copies the logical structure faithfully from the logic $\mathbb{L}(H)$ to $\mathbb{P}(H)$.
C. Since the sequence is pairwise orthogonal $\bigvee_{k=1}^{n} P_k = \sum_{k=1}^{n} P_k \leqq I_H$ for all $n \in \mathbb{N}$. This shows that the sequence of projections is summable. The equality then follows immediately from Theorem 5B.16.

5B.19. Theorem

(i) If $PQ = QP$, then PQ is Hermitian by Theorem 5B.5 and idempotent because $PQPQ = P^2Q^2 = PQ$. Conversely, if PQ is idempotent and Hermitian, then $PQ = QP$ by Theorem 5B.5.
(ii) $P[Q[H]] \subseteq P[H]$, and $PQ[H] = Q[P[H]] \subseteq Q[H]$, so $PQ[H] \subseteq$

$P[H] \cap Q[H]$. Conversely, if $x \in Q[H]$, $Qx = x$; and if $x = Qx \in P[H]$, then $x = Px = PQx \in PQ[H]$. This shows that $P[H] \cap Q[H] \subseteq PQ[H]$.

5B.20. Theorem

Suppose $PQ = QP$. Clearly $(K \cap L) \vee (K \cap L^{\perp}) \subseteq K$. Conversely, suppose $x \in K$. Then $x = Qx + (I - Q)x$ by Theorem 5B.8(v) and Theorem 4.23 (the finite projection theorem). Thus, $x = Px = PQx + P(I - Q)x$. By Theorem 5B.19, therefore, $PQx \in K \cap L$, and $P(I - Q)x \in K \cap L^{\perp}$. Consequently, $x \in (K \cap L) \vee (K \cap L^{\perp})$. This establishes the compatibility of K and L by Theorem 3B.16.

If K and L are compatible, there exist pairwise orthogonal subspaces $R, S, T \in \mathbb{L}(H)$ with $R \vee S = R + S = K$, and $R \vee T = R + T = L$. Let $R^{\sim}, S^{\sim}, T^{\sim}$ be the projections of H onto R, S, T, respectively. Then by Theorem 5B.16 $P = R^{\sim} + S^{\sim}$ and $Q = R^{\sim} + T^{\sim}$. Now by Theorem 5B.9 $PQ = (R^{\sim} + S^{\sim})(R^{\sim} + T^{\sim}) = R^{\sim 2} + R^{\sim}T^{\sim} + S^{\sim}R^{\sim} + S^{\sim}T^{\sim} = R^{\sim 2} = R^{\sim}$. Similarly, $QP = (R^{\sim} + T^{\sim})(R^{\sim} + S^{\sim}) = R^{\sim 2} = R^{\sim}$, and the proof of the first sentence in the theorem is complete.

Since the logic structure of $\mathbb{P}(H)$ is copied from $\mathbb{L}(H)$, it is obvious that the second sentence is equivalent to the first.

Chapter 6A

6A.2. Lemma

This lemma follows immediately from Theorem 4.12 and the bijection between $\mathbb{P}(H)$ and $\mathbb{L}(H)$.

6A.5. Examples

A. If $x, y \in [a, b]$ and $t \in \mathfrak{R}$ with $0 \leqq t \leqq 1$, then $a \leqq tx + (1 - t)y \leqq b$.
B. A simple sketch should convince you that this is obvious.
C. If $x, y \in B_H^r$ and $t \in \mathfrak{R}$ with $0 \leqq t \leqq 1$, then $\| tx + (1 - t)y \| \leqq t \| x \| + (1 - t) \| y \| \leqq tr + (1 - t)r = r$.
D. The point $\frac{1}{3}(1, 0) + \frac{2}{3}(0, 1)$ is not a member of D.
E. It is obvious that for all $x, y \in M$ and *all* $t \in \mathfrak{R}$, $tx + (1 - t)y \in M$.

6A.6. Theorem

Suppose $s_1, s_2 \in \Omega_\sigma$ and $t \in \mathfrak{R}$ with $0 \leqq t \leqq 1$. For all $p \in L$ $(ts_1 + (1 - t)s_2)(p) = ts_1(p) + (1 - t)s_2(p) \in [0, 1]$, because $[0, 1]$ is convex. It is easy to establish the remaining properties to show that $ts_1 + (1 - t)s_2$ is a σ-additive state on L.

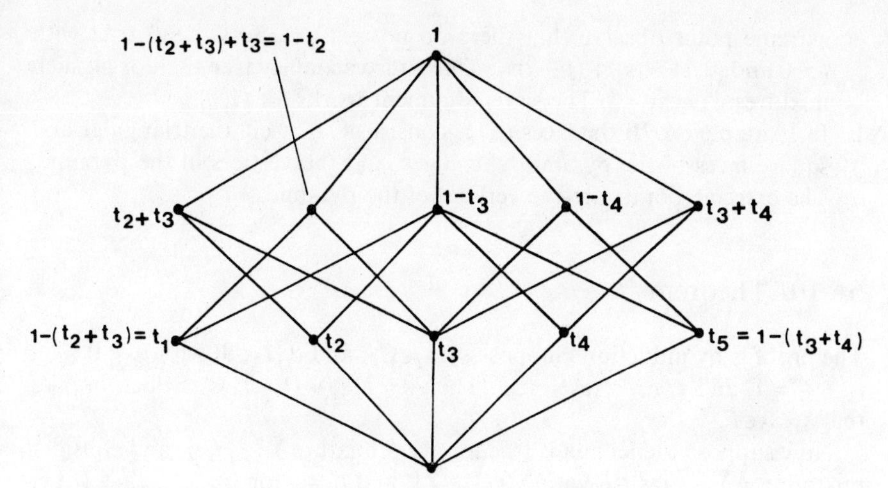

Figure CO6A.1a. Values of a state at all outcomes of the bowtie logic once values t_2, t_3 and t_4 are determined.

6A.7B. Example

See Figure CO6A.1a and b.

6A.9 Examples

A. In Example 6A.7A there is only one face that is not an extreme point: Ω_σ itself. There are two extreme points, $(0, 1)$ and $(1, 0)$. To see that $(0, 1)$ is an

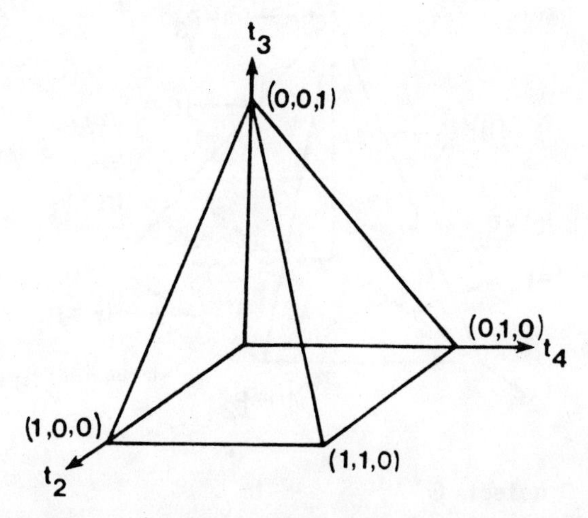

Figure CO6A.1b. The set Ω_σ consists of all points (t_2, t_4, t_3) in the solid pyramid shown.

extreme point observe that there do not exist $s_1, s_2 \in \Omega_\sigma$ and $t \in \mathfrak{R}$ with $t > 0$ and $(0,1) = ts_1 + (1-t)s_2$. Thus, it is vacuously true that for all such mixtures, $s_1 = s_2 = s$. The same argument works for $(1,0)$.

B. In Example 6A.7B the faces of Ω_σ consist of Ω_σ itself, the triangular and square faces of the pyramid, the edges, and the vertices of the pyramid. The extreme points are the vertices of the pyramid.

6A.10. Theorem

The proof is by induction. Suppose $s_1, s_2 \in \Omega_\sigma$, and $t_1, t_2 \in \mathfrak{R}$ with $t_1 \neq 0 \neq t_2$, $t_1 + t_2 = 1$, and $t_1 s_1 + t_2 s_2 \models s \in F$. Then $s = t_1 s_1 + (1 - t_1) s_2$, which implies that $s_1, s_2 \in F$.

Now suppose the lemma is true for every mixture $\sum_{k=1}^{n} t_k s_k$, and consider mixture $s = \sum_{k=1}^{n+1} t_k s_k \in F$ with $\sum_{k=1}^{n+1} t_k = 1$, and $t_k \neq 0$ for $k = 1, \ldots, n+1$. Let $s_0 = \sum_{k=1}^{n} (t_k/(1 - t_{n+1})) s_k$. Then $s = (1 - t_{n+1}) s_0 + t_{n+1} s_{n+1}$, which implies that $s_0, s_{n+1} \in F$. Further, since $\sum_{k=1}^{n} (t_k/(1 - t_{n+1})) = 1$, $s_k \in F$ for $k = 1, \ldots, n$ by the induction assumption.

6A.14. Example

The detectable properties for the state space in Example 6A.7B are shown in Figure CO6A.2 with the propositions that detect them indicated.

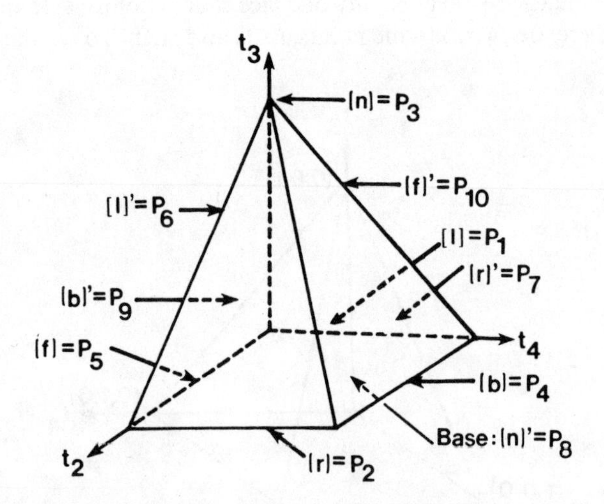

Figure CO6A.2. The detectable properties for the state space of the bowtie manual.

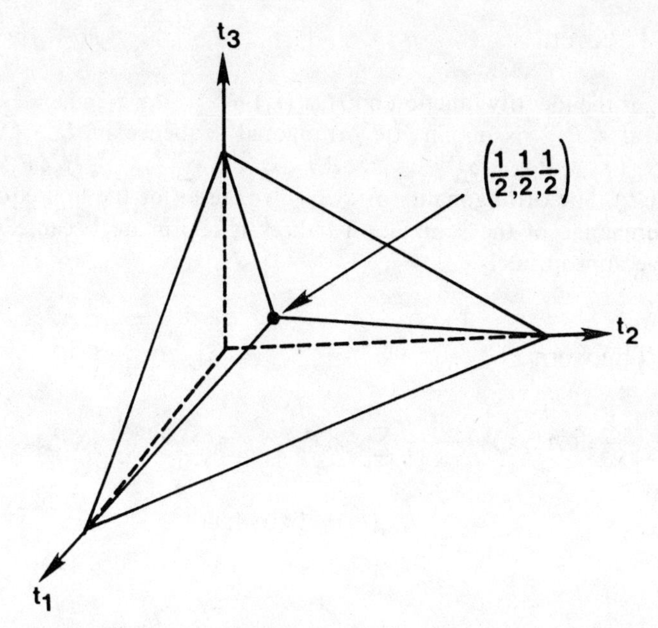

Figure CO6A.3. The weight space for the three-chamber firefly manual.

6A.16. Example

The required sketch appears in Figure CO6A.3.

6A.18A. Theorem

If $\mathbb{K}$ is a pairwise disjoint countable collection of Borel subsets of $\mathfrak{R}$, then $(s\circ\mathcal{O})(\cup\mathbb{K}) = s(\vee_{K\in\mathbb{K}}\mathcal{O}(K)) = \sum_{K\in\mathbb{K}} s\circ\mathcal{O}(K)$, because the sequence $\langle\langle \mathcal{O}(K)\rangle\rangle$ is pairwise orthogonal in L. Further, $s\circ\mathcal{O}(\mathfrak{R}) = s(1_L) = 1$.

Chapter 6B

6B.2. Theorem

Clearly, $s_x(1_L) = s_x(I_H) = (1/\|x\|^2)\langle x, x\rangle = 1$, where I_H is the identity function on H. If $\langle\langle P_k\rangle\rangle$ is a pairwise orthogonal sequence of projections, then $s_x(\vee_{k=1}^{\infty} P_k) = s_x(\sum_{k=1}^{\infty} P_k) = (1/\|x\|^2)\langle\sum_{k=1}^{\infty} P_k x, x\rangle = \sum_{k=1}^{\infty}(1/\|x\|^2)\langle P_k x, x\rangle = \sum_{k=1}^{\infty} s_x(P_k)$, the first equality following from Theorem 5B.18C.

6B.4. Theorem

Since 1_L is the identity function on H, $s_y(1_L)=\sum_{k=1}^{\infty} t_k\langle x_k, x_k\rangle=\sum_{k=1}^{\infty} t_k=1$.

Now if $\langle\langle P_j\rangle\rangle$ is a pairwise orthogonal sequence in L, $s_y(\vee_{j\in\mathbb{N}} P_j)=\sum_{k=1}^{\infty} t_k s_{x_k}(\sum_{j=1}^{\infty} P_j)=\sum_{k=1,j=1}^{\infty} t_k\langle P_j x_k, x_k\rangle=\sum_{j=1}^{\infty}(\sum_{k=1}^{\infty} t_k\langle P_j x_k, x_k\rangle)=\sum_{j=1}^{\infty} s_y(P_j)$. The orthogonality of $\langle\langle P_j\rangle\rangle$ was used for the first equality and the interchange of the summation indices is legitimate because the series converge appropriately.

6B.7. Theorem

$$\begin{aligned}\frac{1}{\|x\|^2}\langle Ax, x\rangle &= \frac{1}{\|x\|^2}\sum_{k=1}^{\infty}\langle\lambda_k P_k x, x\rangle=\sum_{k=1}^{\infty}\lambda_k\frac{1}{\|x\|^2}\langle P_k x, x\rangle \\ &= \sum_{k=1}^{\infty}\lambda_k s_x(P_k)=\operatorname{Exp}(A, x).\end{aligned}$$

6B.9. Lemma

First, $\mathcal{O}_A(\mathfrak{R})=\vee_{\lambda_k\in\mathfrak{R}} P_k=1_L$. Next suppose $\langle\langle T_j\rangle\rangle$ is a pairwise disjoint sequence of subsets of $\mathfrak{R}$ and let $W=\bigcup_{j=1}^{\infty} T_j$. Then $\langle\langle\mathcal{O}_A(T_j)\rangle\rangle=\langle\langle\vee_{\lambda_k\in T_j} P_k\rangle\rangle$ is a pairwise orthogonal sequence in L, so $\mathcal{O}_A(W)=\vee_{\lambda_k\in W} P_k=\vee_{j=1}^{\infty}\mathcal{O}_A(T_j)$.

6B.10. Theorem

Let I be the identity function on H, let $\mu=s_x\circ\mathcal{O}_A$, and let $E=\{\lambda_k\mid k\in\mathbb{N}\}$. Then $\operatorname{Exp}_{\mathcal{O}_A}(s_x)=\int_{\mathfrak{R}} I\,d\mu$.

We consider first the case $\lambda_k\geqq 0$ for all $k\in\mathbb{N}$. Observe that for all $r\in\mathfrak{R}$, if $r\notin E$, then $\mu\{r\}=0$. Since E is countable, this implies that I is equal μ-almost everywhere to the function

$$f(t)=\begin{cases}0, & \text{if } t\in\mathfrak{R}\backslash E,\\ \lambda_k, & \text{if } t=\lambda_k\in E.\end{cases}$$

Then by Theorem 1B.7 (the monotone convergence theorem) we have $\int_{\mathfrak{R}} I\,d\mu=\int_{\mathfrak{R}} f\,d\mu=\lim_{n\to\infty}\langle\langle\int_{\mathfrak{R}} h_n\,d\mu\rangle\rangle$, where the simple functions h_n are defined for each $n\in\mathbb{N}$ by

$$h_n(t)=\begin{cases}\lambda_k & \text{if } t=\lambda_k\in\{\lambda_1,\dots,\lambda_n\},\\ 0 & \text{otherwise.}\end{cases}$$

But then for every $n\in\mathbb{N}$ $\int_{\mathfrak{R}} h_n\,d\mu=\sum_{k=1}^{n}\lambda_k\mu\{\lambda_k\}=\sum_{k=1}^{n}\lambda_k s_x\circ\mathcal{O}\{\lambda_k\}=\sum_{k=1}^{n}\lambda_k(1/\|x\|^2)\langle P_k x, x\rangle=(1/\|x\|^2)\langle\sum_{k=1}^{n}\lambda_k P_k x, x\rangle$. Taking the limit of both sides as n goes to infinity completes the proof for this case.

In case $\lambda_k < 0$ for some $k \in \mathbb{N}$, we write $E_1 = \{\lambda_k \mid \lambda_k \geqq 0\}$ and $E_2 = \{\lambda_k \mid \lambda_k < 0\}$ and observe that

$$\int_{\Re} f \, d\mu = \int_{E_1 \cup E_2} f^+ \, d\mu - \int_{E_1 \cup E_2} f^- \, d\mu = \int_{E_1} f^+ \, d\mu - \int_{E_2} f^- \, d\mu.$$

Then we apply the proof above to both integrals on the right.

Chapter 7A

7A.4. Lemmas

Let λ be an eigenvalue for operator A on H.

A. There is a nonzero vector x in H with $Ax = \lambda x$, so $S(A, \lambda) \neq \{0\}$. A routine calculation shows that $S(A, \lambda)$ is a linear manifold.

B. Justify each step in the following string of equalities. The proof then follows easily. Suppose $0 \neq x \in S(A, \lambda)$. Then $\lambda\langle x, x\rangle = \langle \lambda x, x\rangle = \langle Ax, x\rangle = \langle x, Ax\rangle = \langle x, \lambda x\rangle = \lambda^*\langle x, x\rangle$, which shows that $(\lambda - \lambda^*)\langle x, x\rangle = 0$ and hence that $\lambda = \lambda^*$.

C. Suppose $\lambda_1, \lambda_2 \in \pi(A)$ and $\lambda_1 \neq \lambda_2$. Suppose $x \in S(A, \lambda_1)$ and $y \in S(A, \lambda_2)$. Justify each step in the following string of equalities: $\lambda_1\langle x, y\rangle = \langle \lambda_1 x, y\rangle = \langle Ax, y\rangle = \langle x, Ay\rangle = \langle x, \lambda_2 y\rangle = \lambda_2^*\langle x, y\rangle = \lambda_2\langle x, y\rangle$. Thus $(\lambda_1 - \lambda_2)\langle x, y\rangle = 0$, so $\langle x, y\rangle = 0$.

7A.6. Examples

The task is to find complex numbers λ such that $\det(A - \lambda I_H) = 0$. The point of this project is to familiarize the reader with a few basic examples, not to serve as a complete lesson in finding eigenvalues in complicated situations, a nontrivial problem. So, while there are many sophisticated techniques for finding eigenvalues, we use a straightforward approach on these matrices, which were chosen for their simplicity.

For the first matrix, we write

$$\det\begin{bmatrix} 1-\lambda & 2 \\ -8 & 11-\lambda \end{bmatrix} = (1-\lambda)(11-\lambda) + 16 = (\lambda - 9)(\lambda - 3).$$

Thus, two eigenvalues are $\lambda_1 = 9$ and $\lambda_2 = 3$. Now a vector $\begin{bmatrix} a_1 \\ b_1 \end{bmatrix}$ is in $S(A, 9)$ if and only it satisfies

$$\begin{bmatrix} 1-9 & 2 \\ -8 & 11-9 \end{bmatrix}\begin{bmatrix} a_1 \\ b_1 \end{bmatrix} = \begin{bmatrix} 0 \\ 0 \end{bmatrix}. \tag{*}$$

The solution set for the system of simultaneous linear equations corresponding to $(*)$ is

$$\left\{\begin{bmatrix} a_1 \\ b_1 \end{bmatrix} \middle| b_1 = 4a_1\right\} = \left\{\alpha\begin{bmatrix} 1 \\ 4 \end{bmatrix} \middle| \alpha \in \mathfrak{C}\right\}.$$

Thus $S(A, 9)$ is a one-dimensional subspace of H.

Calculations similar to those just used show that $S(A, 3)$ is the one-dimensional subspace of H spanned by $\left\{\begin{bmatrix} 1 \\ 1 \end{bmatrix}\right\}$.

The second matrix has eigenvalues i and 1, and by calculations similar to those used for the first matrix, it is not difficult to show that

$$S(A, i) = \left\{\alpha\begin{bmatrix} 1 \\ 0 \end{bmatrix} \middle| \alpha \in \mathfrak{C}\right\} \quad \text{and} \quad S(A, 1) = \left\{\alpha\begin{bmatrix} 0 \\ 1 \end{bmatrix} \middle| \alpha \in \mathfrak{C}\right\}.$$

In the third matrix, we have $\det(A - \lambda I_H) = \lambda^2 - 2a\lambda + a^2 + b^2$. From the quadratic formula, $\lambda = (2a \pm \sqrt{(4a^2 - 4(a^2 + b^2))})/2$, so that $\lambda_1 = a + ib$ and $\lambda_2 = a - ib$ are two eigenvalues. Calculations similar to those used for earlier matrices lead to

$$S(A, \lambda_1) = \left\{\alpha\begin{bmatrix} 1 \\ i \end{bmatrix} \middle| \alpha \in \mathfrak{C}\right\} \quad \text{and} \quad S(A, \lambda_2) = \left\{\alpha\begin{bmatrix} 1 \\ -i \end{bmatrix} \middle| \alpha \in \mathfrak{C}\right\}.$$

The only eigenvalue for the fourth matrix is $\lambda = 3$, and $S(A, 3) = H$.

7A.7. Theorem

The induction proof is a bit trickier than one might at first suspect. Let us begin naively.

Let n be the dimension of H. If $n = 1$, let $S_1 = H$, and (i) and (ii) follow immediately.

Suppose now that the theorem is true for all spaces of dimension n or less, and consider space H of dimension $n + 1$ and operator A on H. By Theorem 7A.5, there exist λ in $\pi(A)$ and nonzero vector x in H with $Ax = \lambda x$. Let $S_1 = \text{span}\{x\}$, and let $T = S_1^{\perp}$. Then $\dim(T) = n$. It is time now to invoke the induction assumption on T, but we must be careful. To invoke the induction assumption, we need an operator on T. To use the most likely candidate, $A|T$, we must show $A[T] \subseteq T$, but this will not generally be true. From Theorem 5B.3 we know that since S_1 is invariant under A, the space T is invariant under $A^{\#}$. We might try then using the induction assumption on T and the operator $A^{\#}|T$ on T. That doesn't work either, although it might be instructive for you to try it to see the motivation for the following maneuver.

Let us amend our second sentence in the preceding paragraph by stating instead that there exist λ in $\pi(A^{\#})$ and nonzero vector x in H with $A^{\#}x = \lambda x$.

Now let $S_1 = \text{span}\{x\}$ and let $T = S_1^{\perp}$. Since S_1 is invariant under $A^{\#}$, we know that T is invariant under A. That is, $A|T$ is an operator on T. Now the induction assumption gives us spaces $\{0\} = S_0 \subseteq S_1 \subseteq \cdots \subseteq S_n = T$, where each S_k is of dimension k, and reduces $A|T$, hence reduces A. Letting $S_{n+1} = H$ completes the proof.

7A.8. Corollary

Let A be a linear operator on H. Let $S_0, \ldots, S_n$ be subspaces of H satisfying (i) and (ii) of Theorem 7A.7. Let x_1 be a unit vector in S_1. Then there exists λ_{11} in $\mathfrak{C}$ with $Ax_1 = \lambda_{11}x_1$, because S_1 is of dimension one. Since $x_1 \in S_1 \subseteq S_2$ and $\dim(S_2) = 2$, there is a unit vector x_2 in S_2 with $x_2 \perp x_1$. Continuing in this manner, we can construct orthonormal basis $B_A = \{x_1, \ldots, x_n\}$, which is a basis for $S_n = H$.

Let us find the matrix representation $[[\lambda_{kj}]]_{B_A}$ for A with respect to basis B_A. For integer k with $1 \leqq k \leqq n$, the kth column of the matrix is given by the coefficients in the sum $Ax_k = \sum_{j=1}^{n} \lambda_{kj} x_j$. Since $A[S_k] \subseteq S_k$, we know $Ax_k \in S_k$, so $\lambda_{kj} = 0$ for $k < j \leqq n$. This proves that the matrix is in upper diagonal form.

Now $\det(A - \lambda I_H)$ is independent of which matrix representation is used for $A - \lambda I_H$, and for the matrix with respect to basis B_A, $\det(A - \lambda_{kk} I_H) = 0$ for each $k = 1, \ldots, n$ because the determinant equals the product of the diagonal entries in an upper triangular matrix and the kth diagonal entry in $(A - \lambda_{kk})I_H$ is zero. This proves that each λ_{kk} is an eigenvalue of A and completes the proof of the theorem.

Chapter 7B

7B.2. Theorem

If $A - \lambda I_H$ is not injective, then there is a nonzero vector $x \in H$ with $(A - \lambda I_H)x = 0$, and the constant sequence $\langle\langle x \rangle\rangle$ will satisfy the conclusion of the theorem. If $A - \lambda I_H$ is injective, then by Theorem 5A.5(iv) its inverse is not bounded if and only if there exists a sequence $\langle\langle y_k \rangle\rangle$ of unit vectors with $\lim_{k \to \infty} \|(A - \lambda I_H)^{\leftarrow} y_k\| = \infty$. Then a sequence satisfying the conclusion of the theorem is $\langle\langle x_k \rangle\rangle$ where $x_k = (A - \lambda I_H)^{\leftarrow} y_k / \|(A - \lambda I_H)^{\leftarrow} y_k\|$.

7B.5. Lemma

(i) Since $(L \setminus K) \cap K = \varnothing$, $M(L \setminus K) + M(K) = M(L)$. This proves the equality and also establishes that $M(L) - M(K)$ is a projection. The inequality follows from the fact that the image of $M(K)$ is a subspace of the image of $M(L)$.

(ii) By breaking $K \cup L$ into disjoint parts we obtain $M(K \cup L) = M(K \backslash L) + M(L \backslash K) + M(K \cap L)$. Adding $M(K \cap L)$ to both sides of this equation, we obtain $M(K \cup L) + M(K \cap L) = M(K \backslash L) + M(K \cap L) + M(L \backslash K) + M(K \cap L) = M(K) + M(L)$, the last equality following from the fact that $K \cap L$ is disjoint from both $K \backslash L$ and $L \backslash K$. This proves the first equality.

To prove the second equality, observe from (i) that $M(K \cap L) \leqq M(K) \leqq M(K \cup L)$, whence it follows that $M(K)M(K \cap L) = M(K \cap L)$, and $M(K)M(K \cup L) = M(K)$. If we now multiply (the "multiplication" is function composition) both sides of the first equation in (ii) by $M(K)$, we obtain $M(K) + M(K \cap L) = M(K) + M(K)M(L)$, and this proves the desired equality.

(iii) If $K \cap L = \varnothing$, then $0 = M(K \cap L) = M(K)M(L)$ from (ii). Thus, $M(K) \perp M(L)$ from Theorem 5B.9.

This completes the proof of Lemma 7B.5.

7B.7. Lemma

It is only required to prove that $\mu_{M,x,y}$ is σ-additive. If $\langle\langle K_j \rangle\rangle$ is a pairwise disjoint sequence in $\mathbb{B}$, then $\sum_{j=1}^{\infty} M(K_j) \leqq M(\mathfrak{R}) = I_H$, so the series converges, and we have $\mu_{M,x,y}(\bigcup_{j=1}^{\infty} K_j) = \langle [\sum_{j=1}^{\infty} M(K_j)]x, y \rangle = \sum_{j=1}^{\infty} \langle M(K_j)x, y \rangle = \sum_{j=1}^{\infty} \mu_{M,x,y}(K_j)$. If $x = y$, then for each j, $\langle M(K_j)x, x \rangle$ is a real number because the projection $M(K_j)$ is Hermitian.

7B.9. Lemma

Let A be a Hermitian operator on finite dimensional Hilbert space H, and let M_A be defined as indicated.

(i) Let $\langle\langle K_j \rangle\rangle$ be a pairwise disjoint sequence in $\mathbb{B}$, and let $K = \bigcup_{j=1}^{\infty} K_j$. If $\pi(A) \cap K = \varnothing$, then $M_A(K_j) = 0_H$ for all j, and so $0_H = M_A(K) = \sum_{j=1}^{\infty} M_A(K_j)$. Now since $\pi(A)$ is finite, only finitely many of the terms in $\langle\langle K_j \rangle\rangle$ are mapped by M_A to nonzero projections. It suffices then to show that M_A is pairwise additive. From the spectral resolution theorem we know that for $\lambda \in \pi(A)$, P^{λ} is the projection of H onto $S(A, \lambda)$, so that if $\lambda_1, \lambda_2 \in \pi(A)$, and $\lambda_1 \neq \lambda_2$, then $P^{\lambda_1} \perp P^{\lambda_2}$ by Lemma 7A.4C. Thus if K_{j_1} and K_{j_2} are disjoint Borel sets, they must have no member of $\pi(A)$ in common, and so

$$M_A(K_{j_1}) = \sum_{\lambda \in \pi(A) \cap K_{j_1}} P^{\lambda} \quad \text{and} \quad M_A(K_{j_2}) = \sum_{\lambda \in \pi(A) \cap K_{j_2}} P^{\lambda}$$

are projections that are orthogonal to each other. Then

$$M_A(K_{j_1} \cup K_{j_2}) = \sum_{\lambda \in \pi(A) \cap (K_{j_1} \cup K_{j_2})} P^\lambda = \sum_{\lambda \in \pi(A) \cup K_{j_1}} P^\lambda + \sum_{\lambda \in \pi(A) \cup K_{j_2}} P^\lambda$$
$$= M_A(K_{j_1}) + M_A(K_{j_2}),$$

and this sum is a projection by Theorem 5B.15.

(ii) The first equality follows immediately from the spectral resolution theorem and the additivity of the inner product. For the second equality consider the simple function

$$f(x) = \begin{cases} x & \text{if } x \in \pi(A), \\ 0 & \text{if } x \in \mathfrak{R} \text{ and } x \notin \pi(A), \end{cases}$$

which is $\mu_{M_A,x,y}$-almost everywhere equal to the identity function $I_{\mathfrak{R}}$ on $\mathfrak{R}$. It then follows from Theorem 1B.8 that

$$\int_{\mathfrak{R}} I_{\mathfrak{R}} \, d\mu_{M_A,x,y} = \int_{\mathfrak{R}} f \, d\mu_{M_A,x,y}.$$

The integral on the right is then $\sum_{\lambda \in \pi(A)} f(\lambda)\mu_{M_A,x,y}\{\lambda\} = \sum_{\lambda \in \pi(A)} \lambda \langle M_A\{\lambda\}x, y\rangle = \sum_{\lambda \in \pi(A)} \lambda \langle P^\lambda x, y\rangle$. This completes this proof.

7B.10. Theorem

Suppose M is a spectral measure over Hilbert space H, and f is a $\mathfrak{C}$-valued bounded measurable function on $\mathfrak{R}$.

Let $x \in H$, and define functional φ_x on H by

$$\varphi_x(y) = \int_{\mathfrak{R}} f \, d\mu_{M,x,y} \quad \text{for all } y \text{ in } H.$$

First we shall show that φ_x is a bounded, conjugate-linear functional on H. The domain of φ_x is H, because f is $\mu_{M,x,y}$-integrable over $\mathfrak{R}$ for all y in H. This follows from Theorem 1B.7 because f is bounded, and for all $y \in H$, $\mu_{M,x,y}(\mathfrak{R}) < \infty$. To see that φ_x is bounded let $L = \sup\{|f(t)| \mid t \in \mathfrak{R}\}$. Then by Theorem 1B.7 for all y in H $|\varphi_x(y)| \leqq L|\mu_{M,x,y}(\mathfrak{R})| = L|\langle x, y\rangle| \leqq L\|x\|\,\|y\|$. To see that φ_x is a conjugate-linear functional, suppose K is any Borel set in $\mathbb{B}$ and y_1, $y_2 \in H$ and $\alpha_1, \alpha_2 \in \mathfrak{C}$. Then $\mu_{M,x,\alpha_1 y_1 + \alpha_2 y_2}(K) = \langle M(K)x, \alpha_1 y_1 + \alpha_2 y_2\rangle = \alpha_1^* \langle M(K)x, y_1\rangle + \alpha_2^* \langle M(K)x, y_2\rangle = \alpha_1^* \mu_{M,x,y_1}(K) + \alpha_2^* \mu_{M,x,y_2}(K)$. From this we conclude that

$$\int_{\mathfrak{R}} f \, d\mu_{M,x,\alpha_1 y_1 + \alpha_2 y_2} = \alpha_1^* \int_{\mathfrak{R}} f \, d\mu_{M,x,y_1} + \alpha_2^* \int_{\mathfrak{R}} f \, d\mu_{M,x,y_2},$$

which shows that φ_x is conjugate linear.

By Theorem 5A.11, therefore, there exists a unique x_0 in H such that $\langle x_0, y\rangle = \varphi_x(y) = \int_{\mathfrak{R}} f \, d\mu_{M,x,y}$ for all y in H. The map $B: H \to H$ defined by $x \to x_0$ is the required linear operator on H.

7B.12. Theorem

Let A be a bounded Hermitian operator on finite dimensional Hilbert space H, and let f be a complex-valued bounded measurable function on $\mathfrak{R}$. To show the required equality it suffices to show that for all x and y in H,

$$\left\langle\left(\sum_{\lambda\in\pi(A)} f(\lambda)P^{A,\lambda}\right)x, y\right\rangle = \int_{\mathfrak{R}} f\, d\mu_{M_A,x,y}.$$

Let $x, y\in H$. The function f is equal $\mu_{M_A,x,y}$-almost everywhere to the simple function

$$g(t) = \begin{cases} f(t) & \text{if } t\in\pi(A), \\ 0 & \text{if } t\in\mathfrak{R} \text{ and } t\notin\pi(A). \end{cases}$$

Thus, the integral on the right of the preceding equality equals

$$\sum_{\lambda\in\pi(A)} g(\lambda)\mu_{M_A,x,y}\{\lambda\} = \sum_{\lambda\in\pi(A)} f(\lambda)\langle(M_A\{\lambda\})x, y\rangle = \sum_{\lambda\in\pi(A)} \langle(f(\lambda)P^{A,\lambda})x, y\rangle$$
$$= \left\langle\left(\sum_{\lambda\in\pi(A)} f(\lambda)P^{A,\lambda}\right)x, y\right\rangle.$$

7B.14. Lemma

Suppose A is a Hermitian operator on a finite dimensional Hilbert space H, and let $\pi(A) = \{\lambda_1, \ldots, \lambda_r\}$. From the pairwise orthogonality of the projections P^λ one completes a routine calculation to show that $A^2 = \sum_{\lambda\in\pi(A)} \lambda^2 P^\lambda$. Using induction to establish that the last equality is true if the number 2 is replaced by an arbitrary positive integer n, we arrive at the conclusion that for any polynomial function F, $F(A) = \sum_{\lambda\in\pi(A)} F(\lambda)\, P^\lambda$.

Let $\lambda_k\in\pi(A)$ and consider the polynomial function

$$F_{\lambda_k}(t) = \prod_{\substack{j=1 \\ j\neq k}}^{r} \left[\frac{t-\lambda_j}{\lambda_j-\lambda_k}\right]$$

and the linear operator

$$B = F_{\lambda_k}(A) = \prod_{\substack{j=1 \\ j\neq k}}^{r} \left[\frac{A-\lambda_j I_H}{\lambda_k-\lambda_j}\right].$$

That each coefficient is real follows from Theorem 7A.4B. We shall show that $B = P^{\lambda_k}$.

Let $x\in H$. From Theorem 7A.11 $x = I_H x = \sum_{j=1}^{r} P^{\lambda_j}x$. Let us apply B to this sum. First, consider integer j with $1\leq j\leq r$ and $j\neq k$. Now $AP^{\lambda_j}x = \lambda_j P^{\lambda_j}x$, so $(A-\lambda_j I_H)P^{\lambda_j}x = 0$, and thus $BP^{\lambda_j}x = 0$. Second, if $j\neq k$, then $(A-\lambda_j I_H)P^{\lambda_k}x = \lambda_k P^{\lambda_k}x - \lambda_j P^{\lambda_k}x = (\lambda_k-\lambda_j)P^{\lambda_k}x$, and hence $BP^{\lambda_k}x = P^{\lambda_k}x$. Our first and second calculations applied to the sum by which we have expressed x show that $Bx = P^{\lambda_k}x$, and our proof is complete.

7B.17. Lemma

Let A, B, C, h, f_A, and f_B be as indicated. A rather lengthy but routine calculation using the definition of C shows that $\langle Cx, y\rangle = \langle x, Cy\rangle$ for all x and y in H. One step in the calculation uses the fact that h is real-valued, and another step requires the use of Corollary 7B.16 to establish that $\langle x, P^{B,\gamma} P^{A,\lambda} y\rangle = \langle x, P^{A,\lambda} P^{B,\gamma} y\rangle$ for all λ in $\pi(A)$ and all γ in $\pi(B)$.

To show that $A = f_A(C)$ suppose $\pi(B) = \{\gamma_1, \ldots, \gamma_s\}$. By the definitions of f_A and C we have

$$\begin{aligned} f_A(C) &= \sum_{\substack{\lambda\in\pi(A)\\ \gamma\in\pi(B)}} f_A(h(\lambda,\gamma))P^{A,\lambda}P^{B,\gamma} \\ &= \sum_{\lambda\in\pi(A)} \lambda P^{A,\lambda}P^{B,\gamma_1} + \cdots + \sum_{\lambda\in\pi(A)} \lambda P^{A,\lambda}P^{B,\gamma_s} \\ &= (P^{B,\gamma_1} + \cdots + P^{B,\gamma_s}) \sum_{\lambda\in\pi(A)} \lambda P^{A,\lambda} = I_H A = A. \end{aligned}$$

A similar calculation shows that $B = f_B(C)$, and the proof is complete.

References

The references below do not include many of the well-known books with the words "quantum mechanics" or "Hilbert space" in their titles. Instead they include references for some of the more specialized topics in this book. The references are grouped by subject matter and level of difficulty.

Measure and Integration

Undergraduate Level

Cohn, D., *Measure Theory*, Birkhäuser, Boston, Mass., 1980
Craven, B., *Lebesgue Integration and Measure*, Pitman, Boston, Mass., 1982
Weir, A. J., *Lebesgue Integration and Measure*, Cambridge Univ. Press, London and New York, 1973

Hilbert Space

Undergraduate Level

Halmost, P., *Finite-Dimensional Vector Spaces*, 2nd ed., Van Nostrand, Princeton, N.J., 1958

Graduate Level

Halmos, P., *Introduction to Hilbert Space and the Theory of Spectral Multiplicity*, Chelsea, New York, 1951

Helmberg, G., *Introduction to Spectral Theory in Hilbert Space*, North Holland, Amsterdam, 1969

Quantum Mechanics

Undergraduate Level

Jammer, M., *The Philosophy of Quantum Mechanics*, Wiley, New York, 1974
McWeeny, R., *Spins in Chemistry*, Academic Press, New York, 1970
Mermin, N. D., Quantum Mysteries for Anyone, *Journal of Philosophy*, Vol. LXXVIII, No. 7, July 1981, p. 397
———, Is the Moon There When Nobody Looks? *Physics Today*, April 1985, p. 38
Merzbacher, E., Quantum Mechanics, John Wiley and Sons, New York, 1961

Advanced Level

Einstein, A., Podolsky, B., Rosen, N., Can Quantum-Mechanical Description of Physical Reality be Considered Complete? *Physical Review*, Vol. 47, May 1935, p. 777
Foulis, D., Piron, C., Randall, C. H., Realism, Operationalism and Quantum Mechanics, *Foundations of Physics*, Vol. 13, 1983, p. 813
Hooker, C. A. (ed.), *The Logico-Algebraic Approach to Quantum Mechanics*, Reidel, Dordrecht, Holland/Boston, Mass., 1975
Jauch, J., *Foundations of Quantum Mechanics*, Addison-Wesley, Reading, Mass., 1968
Randall, C. H., Foulis, D., The Operational Approach to Quantum Mechanics, in *Physical Theory as Logico-Operational Structure*, Hooker, C. A. (ed.), Reidel, Dordrecht, Holland, 1978, p. 167
Varadarajan, V. S., *Geometry of Quantum Theory*, Van Nostrand-Reinhold, New York, 1970

Index of Definitions